DRUG DISCOVERY AND TRADITIONAL CHINESE MEDICINE

Science, Regulation, and Globalization

DRUG DISCOVERY AND TRADITIONAL CHINESE MEDICINE
Science, Regulation, and Globalization

edited by

Yuan Lin
Institute for Global Chinese Affairs
University of Maryland

KLUWER ACADEMIC PUBLISHERS
Boston / Dordrecht / London

Distributors for North, Central and South America:
Kluwer Academic Publishers
101 Philip Drive
Assinippi Park
Norwell, Massachusetts 02061 USA
Telephone (781) 871-6600
Fax (781) 681-9045
E-Mail <kluwer@wkap.com>

Distributors for all other countries:
Kluwer Academic Publishers Group
Distribution Centre
Post Office Box 322
3300 AH Dordrecht, THE NETHERLANDS
Telephone 31 78 6392 392
Fax 31 78 6546 474
E-Mail <services@wkap.nl>

Electronic Services <http://www.wkap.nl>

Library of Congress Cataloging-in-Publication Data

Lin, Yuan, 1943-
    Drug discovery and traditional Chinese medicine: science,
Regulation, and globalization / edited by Yuan Lin.
        p. cm.
Includes bibliographical references and index.
    ISBN 0-7923-7411-8 (alk. paper)
    1. Medicine, Chinese –Research—Congresses. 2.
Pharmacology—China—Research—Congresses. I. Title.
R852 .L55 2001
615'.1'0951—dc21

2001029932

TABLE OF CONTENTS

PREFACE..ix

ACKNOWLEDGEMENTS...xi

1. APPROACHING TRADITIONAL CHINESE MEDICINE:
   INHERITANCE AND EXPLORATION
   YONGZHENG HUI 1

2. FLORAS, PLANT CONSERVATION AND CHINA'S
   FUTURE
   PETER H. RAVEN 13

3. NATURAL PRODUCTS DRUG DISCOVERY AND
   DEVELOPMENT AT THE UNITED STATES
   NATIONAL CANCER INSTITUTE
   GORDON M. CRAGG AND DAVID J. NEWMAN 19

4. THE GLOBAL IMPORTANCE OF PLANTS AS SOURCE
   OF MEDICINES AND THE FUTURE POTENTIAL OF
   CHINESE PLANTS
   JAMES S. MILLER 33

5. FOOD, MEDICINAL PLANTS, AND OTHER EDIBLE
   MATERIALS AS SOURCES OF BIOACTIVE COMPOUNDS
   THAT ENHANCE METABOLIC FITNESS AND IMPROVE
   HEALTH
   STEVEN C. BOBZIN AND PAUL BURN.....................................43

6. REGIONS IN CHINA RICH IN RESOURCE FOR
   MEDICINAL PLANTS...55
   A. ANHUI PROVINCE
      ZUOJUN JIANG
   B. HUBIE PROVINCE
      LINCHENG MA

7. THE CAMTOTHECIN EXPIENCE: FROM CHINESE MEDICINAL PLANTS TO POTENT ANTI-CANCER DRUGS
    STRINGNER S. YANG, GORDON M. CRAGG AND DAVID J. NEWMAN..61

8. APPROACHES FOR EVALUATION IMMUNE-MODULATING AND ANTI-TUMOR BIOACTIVITIES IN CHINESE MEDICINAL HERBAL EXTRACTS
    PEI-FEN SU, SHENG-YANG WANG, CHIH-CHIEN HSU, SHOW-JANE SUN, YUAN LIN, PEI-LING KANG, CHIN-JIN LI, JOANNA LIANG, LIE- FEN SHYUR, AND NING-SUN YANG....................................75

9. USING TRANSCRIPTION FACTOR BASED ASSAYS TO STUDY HERBAL PRODUCTS
    DAVID PASCO 83

10. MOLECULAR BASIS FOR MEDICINAL ACTIONS OF ANDROGENS AND GREEN TEA EPIGALLOCATECHIN GALLATE
    SHUTSUNG LIAO, YUNG-HSI KAO, MAI T. DANG, CHING SONG, JUNICHI FUKUCHI, JOHN M. KOKONTIS, AND RICHARD A. HIIPAKKA.. 89

11. STUDIES ON CHEMICAL COMPONENTS AND THEIR PHARMACOLOGICAL ACTIVITIES OF *PANAX GINSENG* ROOT
    LI-XIANGAO, LI XIANG AND LEI JUN..97

12. ANTIBACTERIAL SYNERGY IN RUBRICINE: AN EXTRACT FROM THE ROOTS OF *ARNEBIA EUCHROMA* A CHINESE MEDICINAL HERB
    SPENCER A BENSON, BRIAN P. HIGGINS, CHI S. CHAE, AND YUAN LIN 111

13. ON THE QUALITY ASSESSMENT OF CHINESE PATENT MEDICINE
    PEISHAN XIE AND YUZHEN YAN... 125

14. MODERNIZATION OF TRADITIONAL CHINESE MEDICINE NEEDS FIVE FINGER MOUNTAIN AND GOLDEN HEAD RING
PAUL PUI-HAY BUT. 137

15. A PRACTICAL COMPREHENSIVE APPROACH TO CHINESE MEDICINE RESEARCH
P. C. LEUNG AND, K. P. FUNG...145

16. BACK TO NATURE: THE ALTERNATIVE PARADIGM FOR DRUG DEVELOPMENT
JASJIT S. BINDRA, FRANK C. SCIAVOLINO, DAVID B. MACLEAN, PAUL A. ARMOND, AND PIERRE G. ETIENNE.......................................151

17. GLOBAL MARKET FOR BOTANICAL PRODUCTS
PEGGY BREVOORT 157

18. OPPORTUNITIES AND CHALLENGES OF DEVELOPING MEDICINAL HERBS
MICHAEL CHANG 169

19. TRADITIONAL CHINESE MEDICINES: REGULATORY AND SCIENTIFIC CHALLENGES
MARK EDGAR 177

20. MARCO POLO TECHNOLOGIES: A MODEL FOR A MODERN TCM COMPANY
ROBERT YUAN 191

21. COMMERCIALIZATION OF CHINESE HERBAL MEDICINE IN THE GLOBAL MARKET
CHUNG GUANG SHEN 201

22. THE DEVELOPMENT OF MODERN TCM IN CHINA TIANFA GROUP
JIALONG GONG 209

FDA FORUM
*YUAN YUAN CHIU, CHI-WAN CHEN, LING CHIN, SAMUEL W. PAGE,
LORI LOVE, JOSEPH M. BETZ AND FREDDIE ANN HOFFMAN*.................…....213

PROGRAM..235

INVITED SPEAKERS...241

SUBJECT INDEX..247

PREFACE

The *"First International Conference on Traditional Chinese Medicine: Science, Regulation and Globalization"* was held from August 30 to September 2, 2000 at the University of Maryland at College Park, Maryland. There were approximately 250 participants from the Peoples Republic of China, Taiwan, Hong Kong and the United States. This objective of this conference was to promote international collaboration for the modernization of Traditional Chinese herbal medicines (TCM) and their introduction into the global health care system. It was mainly sponsored by the Ministry of Science and Technology of the People's Republic of China and the NIH National Center for Complementary and Alternative Medicine (NCCAM). It was organized by Dr. William Tai, then director of the Institute of Global Chinese Affairs at the University of Maryland and Dr. Yuan Lin, president of Marco Polo Technologies, Bethesda, MD.

This conference was conceived by Dr. Tai two years earlier recognizing that this was an appropriate time and also the unique location of the University of Maryland. Today, there is a growing recognition of the importance of alternative medicine in modern societies and the rapid loss of knowledge about traditional methods for the treatment of the multitude of human illnesses found throughout the world. TCM has been in common use in China for thousands of years; and many of its formulations are well defined. They are used alongside Western medicine with great success in the world's most populous nation and can be the most important contribution to the world's healthcare system for the 21st century if integrated into modern medical practices.

The focus of the conference were:
- Use of science and technology in the standardization and modernization of TCM.
- Scientific and clinical validation of the safety and efficacy of TCM.
- Integration of TCM into mainstream healthcare systems
- Globalization and marketing of TCM.

The broad themes for the conference were set by the two keynote speakers. Dr. Peter Raven, director of the Missouri Botanical Garden and former Home Secretary of the U. S. National Academy of Sciences, spoke on the flora of China. This consists of 30,000 plant species that represent 10% of the total number in the world. Biodiversity in Chinese plant species has resulted from the existence of a variety of climactic regions ranging from temperate to tropical. TCM has a long history of two thousand years, but is increasingly at risk with the disappearance of plant species due to development. This has led to major efforts by the United Nations and the Chinese government to preserve and promote TCM in a sustainable manner.

One possible solution is to cultivate medicinal plants instead of harvesting them from the wild. Such efforts require the creation of a solid foundation of scientific information. The other keynote speaker was Professor Yongzheng Hui, who had just stepped down as Vice Minister, Ministry of Science and Technology, China, and is director of the new Shanghai Innovative Research Center of Traditional Chinese Medicine. He described the current practice of TCM in China and his vision of how it needs to be modernized. In China, the TCM establishment now consists of 2532 hospitals, 30 universities and colleges, 77 research institutes. Altogether they employ some 40,000 people. 12,000 plant species are used (plus animals and minerals) and there are 100,000 published TCM recipes. While there have been some notable scientific successes, e.g., artemisinin for the treatment of malaria, saponins as anti-inflammatory agents, peony for blocking Ca++ channels, there are significant problems with their rational use. Among them are: the lack of scientific validation, poor standardization, limited collaboration and the lack of databases. He proposed a fundamental change from Experience Based Service to a Knowledge Based Industry. This would require a modern integrated system starting with the sourcing of raw plant materials, determination of effective formulations, manufacture and quality control, clinical testing and global marketing.

The main body of the conference was divided into four sessions:
- Natural resources and conservation of medicinal plant species
- Scientific session I: An ancient healing art in a modern world
- Scientific session II: TCM: A new platform for drug discoveries
- Commercialization of TCM in the global market

In addition, a panel of experts on the regulation of herbal products in the United States held a FDA Forum to discuss regulatory issues relating to manufacturing and marketing of herbal products in the US.

The articles in this volume describe various strategies for the development of TCM as drugs or dietary supplements as well as the use of modern scientific methods to understand their mechanism of action. It also deals with issues of regulation and global commercialization.

The general consensus among the participants is that for TCM to achieve its potential in global health care will require a sophisticated integration of traditional knowledge, historical clinical experience and modern technology - the best of East and West.

Yuan Lin
February 2001

ACKNOWLEDGEMENTS

The conference committee sincerely thanks the following organizations for their enthusiastic support and generous financial donations.

Domestic Sponsors

National Institutes of Health
National Center for Complementary & Alternative Medicine

International Sponsors

Department of International Cooperation, Ministry of Science and Technology, Beijing
Beijing Science and Technology Commission, Beijing Municipal People's Government
The People's Government of Jilin Province
The People's Government of Sichuan Province
Peking University
Institute of BioAgricultural Science, Academia Sinica, Taipei
Pharmaceutical Industry Technology and Development Center, Taipei
Hong Kong Economic and Trade Office, Washington, D.C.
Institute of Chinese Medicine, The Chinese University of Hong Kong

Corporate Sponsors

Pfizer, Inc. (Groton, Connecticut)
UBI Asia (Taipei)
Taiwan Sugar Corporation (Taipei)
King Car Food Industrial Co., Ltd. (Taipei)
Chainex Metal Factory Co., Ltd. (Taipei)
Sinphar Pharmaceuticals (Taipei)

The editor also wishes to express her heartfelt thanks to Dr. William Tai for his trust and confidence in embarking on such a daunting challenge. His dedication to the preservation of flora in China and endless efforts in bringing scientists to work together won him long lasting friendship across the oceans. Without such a special relationship, this conference could not have been a successful one. The editor also wishes to express her sincere thanks to Ms. Rebecca McGinnis for her support and dedication. Efforts from other members of the IGCA staff, Frances Jarvis, Yuedong Zhou, Qun Zhang, Baoyan Cheng, Daifeng Han, Liqun Xu and Duy-Khrong Van are also appreciated. Special thanks also go to Katherine Young, Deborah Jurdjevic and Zhencheng Zhang for their assistance in editing many of the chapters of the book. The editor is most grateful to Dr. Robert Yuan, when times are bleak, he is always there to bring in the sunshine.

Chapter 1

APPROACHING TRADITIONAL CHINESE MEDICINE: INHERITANCE AND EXPLORATION

YONGZHENG HUI
Shanghai Innovative Research Center for Traditional Chinese Medicine
Shanghai, People's Republic of China

Abstract: Traditional Chinese Medicine (TCM) is a treasure house for human beings; it has been used in China for thousands of years and has contributed greatly to the growth and thriving of the Chinese Nation. TCM is not simply equal to Natural medicine; it is a collection of natural resources, Chinese culture, historical experience and modern research. Why is there only little success in searching for New Chemical Entity (NCE) from TCM since the last century? Apart from history and culture, not based on clinical practices, ignorance of water-soluble components probably is the answer.

Compared to Western drugs, TCM have some unique advantages, i.e. a rich array of proven medicinal species, remedies and clinical practices, tailor-made medications, guidance by Chinese medical theory, human centered and low cost. But TCM also possess some disadvantages, i.e. too far from modern sciences, no databases, less coordination and cooperation, poor standardization, and formulation types not accepted, which all need to be solved to ensure further development.

Now a very challenging topic for modernization of TCM is how to transfer its character of Experience based Service to a Knowledge based Economy? In order to eventually build up an international business discipline, there is a chain process including many stages from the production of genuine medicinal species (GAP), finding effective recipes (GLP/GCP), quality control and scientific manufacture (GMP) until global sales network (GSP) can be formed. At the same time, it also opens huge business opportunities for start-up companies searching for new leads from TCM with the injection of modern sciences such as pharmaco-genomics, molecular and cell biology, chemical biology, neuroscience, information science and synthetic organic chemistry.

Yuan Lin (ed.), Drug Discovery and Traditional Chinese Medicine: Science, Regulatory and Globalization, 1-12.
©2001 *Kluwer Academic Publishers. Printed in the Netherlands.*

1. INTRODUCTION

The subject matter of this meeting is the science, regulation and globalization of TCM. To reach the end, both inheritance and exploration are needed.

There is a huge increase in the Chinese population from B.C. to 1994. A more careful look, however, reveals that from B.C. to the 16th century, the Chinese population still remains at a very stable level at around 15 millions people despite large numbers of droughts, floods and wars. After the 16th century, the population grew rapidly. In the year 1994, China's population reached 1.2 billion, the country with the largest population in the world. There is an interesting sharp upward turn around 1660, in Qing Dynasty. Another interest observation is the number of plant species that can be used as drug. According to Chinese literatures, before 16th Century, the number of medicinal species is very limited, less than 1000. According to the latest survey in 1990s, there were more than 12,000 species can be used as drugs. Again, there is a turning point at about the same time period (in 1680s) with the increase in population. We can tentatively conclude that maybe the increase of population is partly resulted from an increasing use of traditional Chinese medicine (TCM).

2. CHINESE MEDICINE: ITS USAGE AND RESOURCE IN CHINA

A recent survey was conducted two years ago among 1,543 households in Beijing, Shanghai and Guangzhou. The result shows that 41% of the families believe that TCM and western medicines are equally effective; 31% prefer TCM to western drugs; and 23% prefer western medicines. Although people's attitudes and loyalty differ about TCM, no one says that TCM doesn't work. Realities in both history and modern society reveal that TCM is very important for the Chinese nationality. Someone says that TCM is a natural medicine. It is correct, but it's not complete. In fact, TCM should be a collection of natural resources, Chinese culture, historical experience and scientific research.

The resources for TCM include two parts. One is human resource. China probably is the only country whose medical staff holds TCM and western medical science at equal status. It's not just legal, but equal. China boasts more than 2,500 TCM hospitals, 30 universities and colleges engaging in the studies of TCM, 51 technical schools of TCM, 77 independent research institutes for TCM, and more than 40,000 professional and technical personnel involved in TCM. That is a very big human resource.

The other resource of TCM is the natural resource. TCM is practiced in diverse locations in China, from the eastern monsoon area to the western drought area all the way to Qinghai and Tibet plateau. China also has variety of species that TCM depends on. Medicinal plants are the major components of TCM but it also includes some animals and minerals. Altogether there are 12,000 species nowadays that can be used as drugs in China. A couple of years ago, an encyclopedia of TCM was published in Shanghai, it includes 9,000 medicinal species. China is a nation of multi-nationalities. Every minor nationality, such as Mongolia, Tibetan, Dai and Miao, has its special drugs.

It is important to realize that TCM is not just natural products. There is a comprehensive collection of ancient medical books and records. Books over 1000 years old still are in existence and can be used as guides to treat diseases.

3. THE THEORETICAL BASIS OF TCM AND ITS RECENT HISTORY IN DRUG DISCOVERIES

TCM is an integral part of Chinese classical philosophy. TCM utilizes concepts and theories from traditional Chinese philosophy, for example, the holism, the pattern identification method, and the five elements, which are gold, wood, water, fire and earth. There is a huge historical accumulation of folk remedies and recipes. According to my personal survey, more than 100,000 proved secret recipes have been published. But about the same number have been spread in countryside and were not recorded. The total number of recipes, therefore, could reach 200,000 or even 300,000. These recipes have been applied to clinical practices for thousands of years and they are very important properties for the whole human race. These recipes indicate the effectiveness and toxicological side-reaction of the drug, the ingredients, and the proper dosage. From ancient time, the recipes, spread in extensive areas, are very useful in searching for new drugs. It is important to emphasize that we should never underestimated what our ancestors said.

In a more recent history, there was a nationwide program searching for anti-malaria medicine from TCM. During the 1970s, in China, more than 500 species were identified. Out of which, 5 were found to be good; among them 2 were found to have the best efficacy. Finally, Artemisimin (qing hao su) displayed the best effect. Scientists from the Chinese Academy of Traditional Chinese Medicine, Institutes of Organic Chemistry and Biophysics worked closely to elucidate the structure by chemical and x-ray methods and to confirm the pharmacological action. The structure was finally elucidated. Then a series of analogues were made because the lactone is unstable in the original molecule. Lactone was transformed to methyl alcohol to stabilize the structure. Scientists also found that this compound is very hydrophobic and it can easily permeate into the membrane. In the presence of ferrous ion it can interact with DNA of malaria parasites to destroy the DNA and rid of the malaria (Fig. 1). For normal cells, there is no malaria parasites DNA inside, since mature red cells have no nucleus, therefore, they cannot be killed. For malaria patients, however, there are malaria parasites in the red blood cells and will demonstrate the above-mentioned reaction.

Tracing back in history, around 400 years ago, in the Ming dynasty, Li Shizhen wrote that artemisimin can cure malaria. Most interestingly, a couple of years ago, an ancient piece of silk from Ma Wang Dui was discovered. Using radioisotope-dating procedure the silk was estimated to be over 1000 years old. From the piece of silk, 4 Chinese characters were recognized, meaning, "Artemisimin can stop malaria". It seems like the modern men are not as smart as ancient people. Often times 100 scientists gathered to perform a new task while over1000 years ago ancient people already the answer.

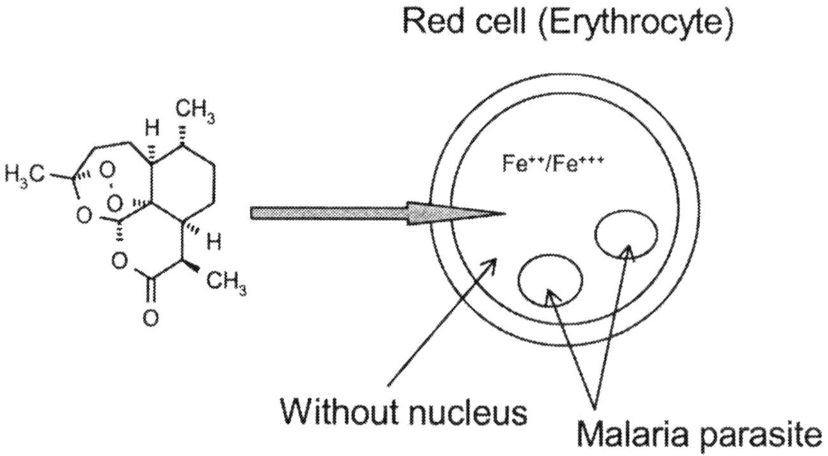

Figure 1. Consideration on the anti-malarial mechanism

The 2nd example is *Achyranthes bibentata* that is a very popular TCM drug nowadays in China. According to Chinese ancient literatures, this specie has two functions. One is "bu qi", which means it is good for health, particularly, for the immune modulation. The other function is the anti-inflammatory effect. In 1988, my group worked with professor Tian and isolated an oligosaccharide from this specie with a molecular weight 1000. It showed very good activity for the immune modulation and for the protection of liver damage. Right now, it is used to treat more than 100 cases of cancer patients. After the chemotherapy and radiotherapy, the immune system of cancer patients is declining. Using this compound, patient's immune system recovered from a very low level of 1000 white cells counts to a normal level of 5000. It has been commercialized as a supplementary drug for cancer patients.

But the compound in *Achyranthes* that has shown an anti-inflammatory effect was not identified, until 1998 when a Japanese scientist, Ida, isolated a family of saponin from this species (Fig. 2). This compound has very high anti-inflammatory activities. The activities are 1000 times higher than sLewis, which is a powerful anti-inflammatory drug. Since the concentration of these compounds in this species is very low. For a long time, no one knows the existence of these families of saponin in the herb, in this herb. My group will be finishing the total synthesis of this family in a couple of months.

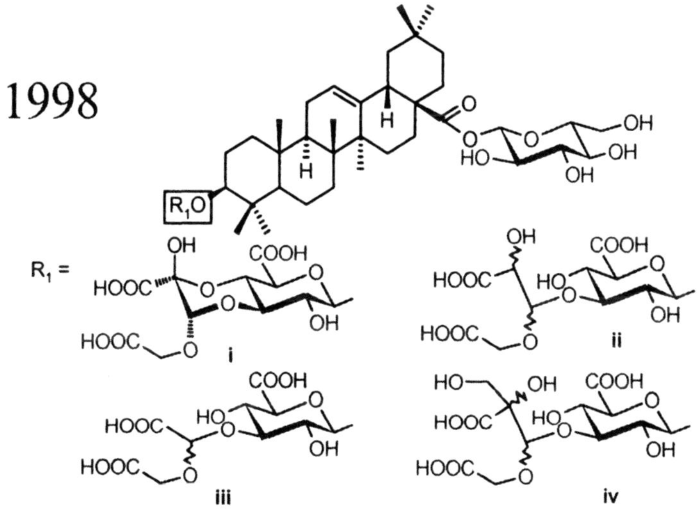

Ida, Y. etal *Bioorg. Med. Chem. Lett.* **1998**, *8*, 2555

From *Achyranthes*

Saponins iii and iv have anti-inflammatory activity 1000-fold potent than sLewis^x

Figure 2. Structure of saponins from Achyranthes

The above are two examples to show that ancient people found a lot of activities and recipes from TCM through clinical paths. They wrote down these findings in books and we should never underestimate them. These books are very useful in searching for new drugs in the modern ages.

4. COMPARISON BETWEEN TCM AND WESTERN DRUGS

Western drugs are often single-targeted. One drug is usually targeted to cure one disease, for instance, anti-bacteria, and anti-virus. Enzymes inhibitors have also become popular these days. TCM, in comparison, is multi-targeted in most cases. For instance, "xiao chai hu tang", soup of Chinese thorowax, can cure chronic hepatitis, kidney diseases, asthma, and skin diseases. This is one difference between western drugs and TCM.

The other one is the opposite ways of searching for new drugs. Western medicine starts with molecular designing, using a the 3 dimensional protein target structure, the genes, the proteins. This will be followed with a combinatorial synthesis to get a large number of compounds. They can be tested on small animals, large animals and finally on human beings. After passing phase 1, phase 2 and phase 3 with volunteers, these compounds finally get the approval from FDA to use it as legal drugs. And they can be used for 50 years after the patent expiration. This is a typical way from molecular designing to a new drug, from animal to human beings.

For TCM however, it is reversed. It starts from previous experience on humans. When someone has a disorder and go to a hospital, a Chinese Traditional Medical doctor will prescribe a recipe and treats the patient as an animal model. If this recipe works then maybe someone will become interested to find out why it works. It then get tested on small animal, which will be followed by molecular level study for why and what. It represents a totally different philosophy from that employed by western doctors.

The 3rd difference between western drugs and TCM lies in inhibition versus regulation. Western medicine is like a killer. It kills the cancer cell, the bacteria, and the virus. Sometimes it is known as inhibitors. But TCM is most likely to play a role as a biological-response modifier (BRM). It interacts first with the normal cell. Then the normal cell releases some molecules to fight the cancel cell. We cooperated with a number of universities in Hong Kong to provide strong evidence in support of the notion that TCM plays the role of BRM rather than inhibitors.

Western medicine works from external stimulation to the local physiological changes. TCM work with the local changes, but it is mostly to human bodies then human bodies fight against the diseases. Traditional medicine pays great attention to environmental changes and the side effects of medicines (See Figure 3 on next page).

It is also worth noting that a Chinese herbal soup of licorice root and Chinese herbaceous peony (1:1) mixture which is an effective sedative. The active components in both species are saponins but the mechanisms of action are different—saponin from peony blocks calcium channel while saponin from licorice root blocks potassium channel. Understanding the actions of different compounds are crucial to drug discovery.

Out of the economic consideration, TCM is much cheaper to develop than western drugs. Then, the question is: "Why there is only so little success in searching for new chemical entities (NCE) from the TCM since the last century?" National Cancer Institute annually searches for a lot of compounds worldwide from plants. But why there's so little success in searching NCE? Maybe texol is a success story, but there is very few. There could be 3 reasons:

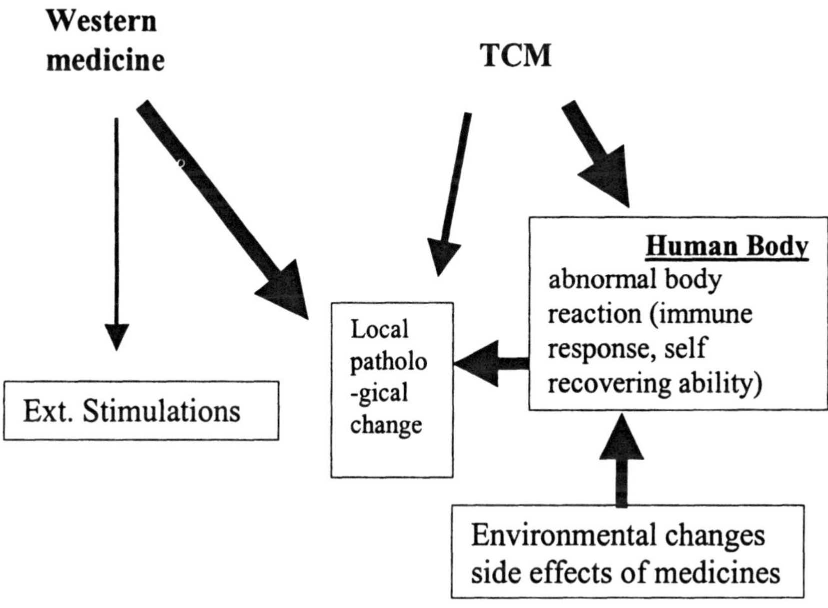

Figure 3. Difference between Western medicine and TCM

First, this kind of searching is away from the history and culture. Just like the *Achyranthes* example, the active compound exist as a very minor component, nobody discovered it. But in the historical literature, it showed anti-inflammatory effect so there must exist molecules with such activities. Therefore if one stays away from the history and culture, one will probably miss a lot of useful compounds.

Secondly, it is not based upon clinical practice. The search is not combined with practical application. If one search for compounds based on a clinically well-established framework, chances are that it would work. Many medical cases have accumulated in 1000 years. Maybe more than 10,000 cased have been tested already. That is the wealth of clinical practices.

The third reason is the ignorance of water-soluble components. Most natural product chemists are searching for new compounds in organic extracts particularly with peculiar structures that are good targets for chemical synthesis. But Chinese people do not drink the acetone extracts; they only drink the water extract. They boil herbs in water and sometimes in dilute alcohol. They never drink extracts of organic solvents. Water-soluble components are very important for TCM and more attention should be paid to them. What kind of products is water-soluble? Beside acid, there are oligo-saccharides, poly-saccharides and saponin. Statistics show that water-soluble components contain 6% of the saponins. Saponin could be a new family of drug, and it deserves more attention.

TCM has 5 unique advantages. The first one is a large number of proven medicinal species, remedies and clinical practices for thousands of years.

The second one is that TCM practices tailor-made medications. The Chinese prescription is based on a one-by-one approach, a customized medication. This characteristic of TCM is very advantageous and it could become a reality as the study on human genes has brought us to the new era.

The third strength comes from the guidance of classical medical theory. In a Chinese hospital, there are many qualified Chinese medical doctor who displays great confidence in providing prescriptions of TCM. They know what kind of species is good a particular individual, which is not so good and which is toxic. Although many classical medical theories is still unknown at his point, nevertheless, they work.

Another important advantage of TCM is that it is human centralized. If one suffers from a stomach disorder, TCM will not only treat the stomach, but rather it treats the whole body. If one suffers from a liver illness, TCM will probably not treat the liver, rather the eyes. TCM treats human body as a whole. This is a very new approach to cure the diseases.

The final advantage is the low cost. Compared with western drugs, it is relatively cheap.

But TCM also has a number of disadvantages. And that's why there is a lot of work remaining for the modernization of TCM. The first weakness of TCM lies in the fact that it is too far away from modern sciences. For most recipes, we don't know what components work and why they work. Even for the ginkgo biloba, maybe the lactone is very important, but it is not known why the lactone is useful for the neuro-system.

The second one is a lack of database. Even with the comprehensive encyclopedia, there is no database like those of the western drugs. Works are in progress to build a real TCM informatics for drug searching.

The third disadvantage of TCM lies in a lack of collaboration among researcher and clinicians. From the literatures there are many recipes. One recipe might have been used to treat 100 patients and the results showed that 6% of them with very good responses, 20% good, and 20% no use. Then another doctor may use the same recipe for another 100 patients. They don't have a cooperative relationship as western doctors do. They employ a family linked, generation-by-generation approach and there is less cooperation among colleagues. The above three disadvantages were observed by a leading scientist, a VP of a multinational pharmaceutical company. He viewed TCM from the perspective of a western scientist. His statements are very reasonable. These are the problems that must be solved.

The fourth weakness is the poor standardization. In most cases, the standardization of TCM is not well established. It is also a long way to go for this end.

Last but not least, formulation type is not user-friendly. A very classical type of formulation is in the shape of ping-pong ball and coated with wax. Due to its big size, it needs to be squeezed into small pieces in order to be swallowed. Another example is the preparation of TCM medical soup. A large number of herbs have to be boiled together. The hours-long process entails attention throughout. Thus, the current formulation is not acceptable and needs to be improved.

5. FROM AN EXPERIENCE-BASED SERVICE TO A KNOWLEDGE-BASED ECONOMY

In the new era, information technology will play an important role in many disciplines of sciences in the coming century. Among the life sciences, TCM will play a very important role. To achieve the objective, traditional TCM has to be transformed to a modern TCM, which stands as a science, service and an industry. Only a world-class modern TCM can turn the traditional experience to a knowledge-based economy. Otherwise there is no future for TCM.

Two factors are very important: one is the novel knowledge creation and the other is the marketing operation. In order to eventually build up an international business discipline, there is a chain process including many stages in modern TCM development (Fig. 4). The production of genuine medicinal species requires a "Good Agricultural Practice" (GAP), finding effective recipes needs "Good Laboratory/Clinical Practice" (GLP/GCP), quality control and scientific manufacture needs "Good Manufacture Practice" (GMP) and global sales network needs "Good Sales Practice" (GSP).

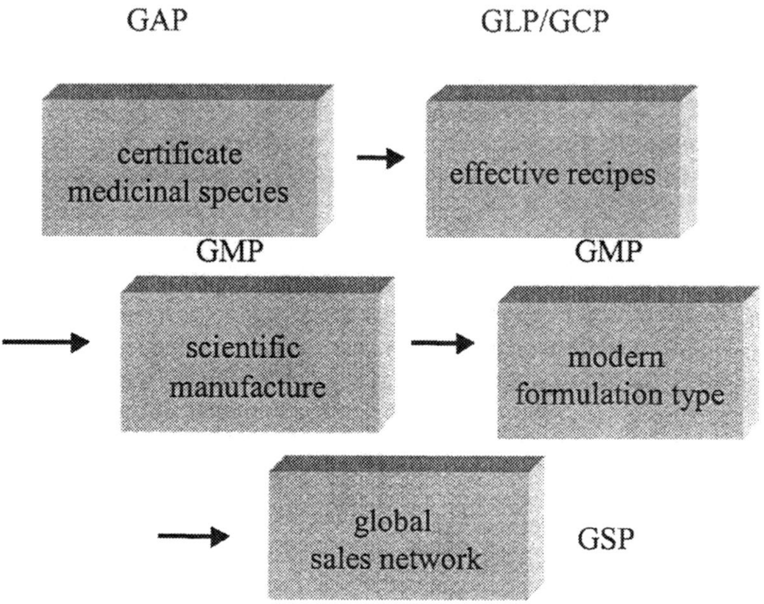

Figure 4. Stages and business opportunities in modern TCM development

There is a need to produce certificates for medicinal species, which means to build up herbal gardens. Right now, there are a lot of herbal gardens in China. The best location to build those herbal gardens is in the Yunnan province (in the southwestern part of China), Wenshan County. It excels all the others in quality. For the modernization of TCM, there is a need to follow the tradition, secure certificate, and build a lot of herbal gardens. The next step is to establish a whole process of quality control for the active components, at least markers. Next step is to determine the contamination such as heavy metals and agricultural chemical residues. Finally, the issuance of certificates for medicinal species, the brand market certificate.

Figure 5 depicts steps leading to the creation of effective recipes. From the record, the collection, and the prescription from traditional Chinese doctors, to the recipe, then to the clinical application, the pharmacology study, the clinical trial, finally to the effective recipe.

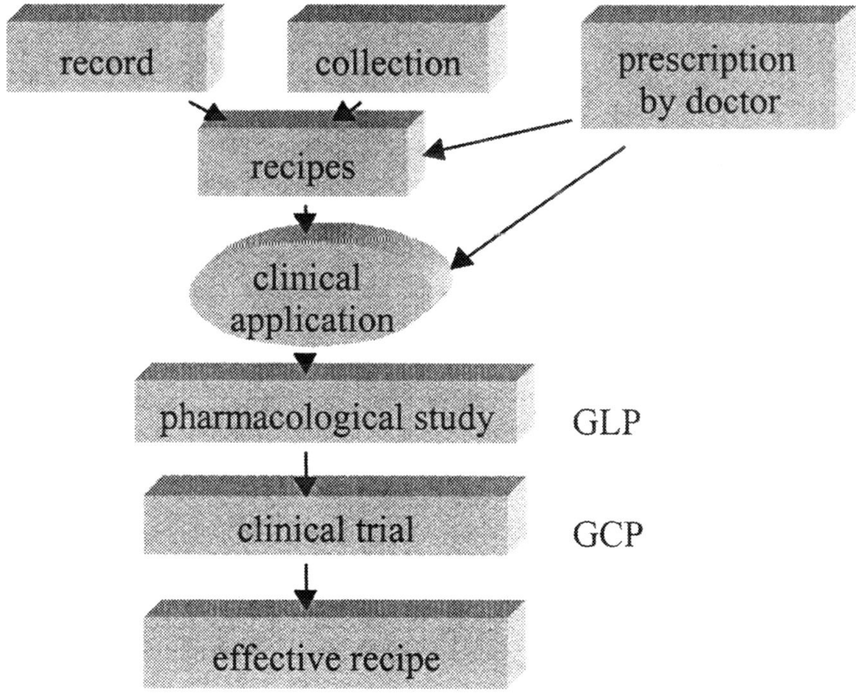

Figure 5. Creation of effective recipes

Figure 6 shows different sales network for TCM products. There is another process to achieve the same purpose. It is from modern TCM to chain stores, to insurance companies, to customer medication, to hospitals, or to patients directly.

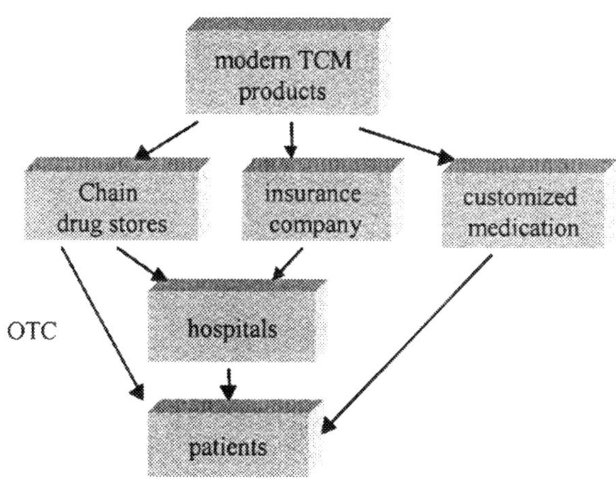

Figure 6. Sales networks for TCM

12

6. CONCLUSION

In summery, there is a bright future for TCM due to the following reasons. First, human genetic project is scheduled to be finished at 2003. It will bring huge opportunities for pharmaceutical industry including TCM. Secondly, pharmaco-genomics is dawning. If one can combine the pharmaco-genomics and TCM, one probably will find a better way of approaching NCE. Thirdly, chemical biology is booming. It studies the interaction between protein (the macro-molecule) and ligand. This is a very good subject in searching for new drugs. Awareness on modern TCM is deepening. From China and outside China, everyone knows TCM works. But the important thing nowadays is how to achieve its modernization. Lastly, venture capitals are beginning to fund life science and health care sectors. They are also interested in TCM. And thus, there is a bright future for TCM!

Chapter 2

FLORAS, PLANT CONSERVATION AND CHINA'S FUTURE

PETER H. RAVEN
Missouri Botanical Garden
St. Louis, MO 63166

1. INTRODUCTION

This chapter will cover four areas: (1) an overview of the Flora of China, its characteristics, and the reasons for its rich development; (2) the formation of databases for an overall understanding of these plants; (3) plant conservation in China; and (4) Chinese medicinal plants and Traditional Chinese Medicine (TCM).

2. AN OVERVIEW OF THE FLORA OF CHINA

To consider the flora of China, it is essential to look at the whole North Temperate Zone, which is a huge important region of floristic richness. China, Europe, and the United States are remarkably close to the same size. In China there are about 30,000 species of vascular plants, about 10 percent of the world total. In the United States, there are about 18,000 species, and in Europe, about 11,000 species. So, the obvious question that needs to be examined in thinking about the plants of China and the Northern Hemisphere generally is "why are there so many more plants in China than in these other two areas?"

The first observation can be made about this is that before the Middle Miocene Period, about 15 million years ago, judging from the fossil record, the assemblages of plants and animals in Europe, in China, and in the United States were about the same, both in numbers of species and in diversity of

Yuan Lin (ed.), Drug Discovery and Traditional Chinese Medicine: Science, Regulatory and Globalization, 13-18.

species. Over the past 15 million years, the climates in these temperate regions have progressively become more seasonal, that is, colder in the winter, hotter in the summer, and drier overall. During that period of time, for some reason, the plants in China survived better than the plants in the other two regions.

It is often assumed that many plants and animal species became extinct during the ice ages in the Pleistocene, because there were big changes during that period of time, locally. But in fact, what various species did, over the past few million years, was not become extinct generally, but rather to move around from one place to the other; to disappear locally, but survive somewhere else. Actually, very few species in the North Temperate region became extinct during the Pleistocene, the last couple of million years, but many disappeared from whole continents over the preceding 15 million years. The most important reason that there is more survival in China seems to be the continuity from tropical to cold temperate regions in Asia, going all the way from South Asia up into the North. That continuity is not matched in either Europe or North America. For example, in Europe, the Alps, the Mediterranean, and the Sahara Desert completely cut off connections of the temperate vegetation with tropical vegetation. And in North America, the Gulf of Mexico, and the deserts of Northern Mexico interrupt the direct connection. So, the richest zone of survival in China of these older relictual species is especially in eastern Sichuan and western Hubei and again in southeastern Yunnan and western Guangxi. Those have been areas that have been relatively favorable for survival for the past 15 million years while the world climate has been changing.

This case can best be illustrated by a few examples, beginning with *Metasequoia. Metasequoia* was abundant all around the Northern Hemisphere from the late Cretaceous onward. Forty million years ago, *Metasequoia* was the commonest tree in the forests of western and northern North America, and was a common forest tree in most or all Northern Hemisphere forests. But it disappeared completely from North America about 15 million years ago, and is represented now by about 6,000 individuals in Sichuan and Hubei, China. Of course, during the past 50 years, it has been widely introduced and there are millions of individuals all around the Northern Hemisphere, all the way to southern Alaska and all across North America and Europe. There are many, many other important relics also of primitive flowering plants that have survived the same way in China, but are known as fossils, both in Europe and in North America. It is worth noting that one of them is *Camptotheca,* which will be discussed in a later chapter.

A second important reason that the numbers of species are so high in China is the obvious one: that there is actually tropical vegetation in China and many species are found only in that tropical vegetation, in southern Yunnan on the mainland, and also offshore. There are about 8,000 species, if not up to 10,000 species, in tropical Yunnan alone, which is a very high proportion of all the plants in China.

A third factor has to do with the plants that occur in middle and higher elevations in all of the mountain ranges in China. As the Indian subcontinent has collided with Asia progressively over the past 50 million years, China's topography has been folded up like a fan with pleats into all the many mountain ranges that go across China, and this has happened mostly over the last 50 million years, and about 40 percent of China is at middle or higher elevations. Much of China in fact is at high elevations, in isolated mountain ranges that have been very important places for the development of large numbers of species of plants in different groups. Consider just a couple of examples: in *Rhododendron*, 650 of the 850 species of the world are found in China; in *Pedicularis*, 350 of the 500 species in the world are in China; and in *Primula*, 300 of the 500 species in the world are in China. At a family level, there also are very high numbers in China, for example: 528 of the 1,000 species of Primulaceae, and 420 of the 700 species of Gentianaceae in the whole world are in China.

There are three reasons then for China's floristic richness: tropical rainforest, unbroken north-south vegetation corridors, and many parallel mountain ranges of recent origin. All of these factors come together to make China a wonderful repository of 10 percent of the plant species in the world, and, by far, the most interesting area in the whole Northern Hemisphere.

3. DATABASES FOR AN OVERALL UNDERSTANDING OF THE FLORA OF CHINA

Two thousand years of writing about the plants of China have greatly increased the knowledge as time has gone by. The Flora of the People's Republic of China, which is a 40-year project soon coming to completion, will provide, in 120 volumes, a comprehensive account of all of the plants of China, in Chinese, with thousands of individual illustrations.

American botanists first began meeting with Chinese botanists in 1979 when there a meeting was held in Berkeley, California, to discuss what kind of projects to be embarked in the future. At that meeting, which Dr. William Tai coordinated and I participated in, Professors Wu Zheng-yi and Yü Te-tsun requested international collaboration in revising the account of the plants of China. The goal was an international standard for dealing with those plants, because a lot of the work was done during the Cultural Revolution and other difficult times when access to specimens, library materials, and other materials in the West was not possible. Meanwhile, studies were going on very rapidly of plants of the Himalayan region, plants of the Asian part of

16

Russia, plants of the Middle Asian regions of the Soviet Union (now separate countries) plants of South Asia, Japan, and so forth. All of these studies needed to be better coordinated with the study of the plants of China, and that is one of the goals of his study.

After many visits to China and many discussions about various issues, by 1987, the project was approved, and Professor Wu Zheng-yi and I became the co-chairs of the Editorial Committee. The plan was to publish 25 volumes of text (to date, there are six volumes in print) and then another 25 volumes of illustrations, of which four have been published so far. This a cooperative program sponsored by the Chinese Academy of Sciences, with the main office in the Institute of Botany in Beijing, but it also involves direct participation by the institutes in Guangzhou, Kunming, and Nanjing. In addition, dozens of individual Chinese scientists in other institutes of the Academy of Sciences and in the universities in China are participating in the cooperative and collaborative accounts of the plants. Outside China, the main cooperating institutions are the Missouri Botanical Garden, the Smithsonian Institution, Harvard University, the California Academy of Sciences, the Royal Botanic Gardens at Kew and Edinburgh, the Natural History Museum in Paris, to some extent, the Natural History Museum in London, and the Komarov Institute in St. Petersburg. The Flora is jointly published by Science Press and the Missouri Botanical Garden, but is completely available on-line, and will be able to be searched for individual words and characters very shortly. It is also continually revised as additional information becomes available, and is related to a computerized checklist and database on all the plants of China. This project was supported by the National Science Foundation in China, the National Science Foundation in the US, as well as other public and private agencies in both countries.

4. CONSERVATION IN CHINA

It is obvious that the tripling in size of the Chinese population since 1949, together with rising levels of affluence and consumption, which are part of economic prosperity, have endangered and threatened many, many plants and animals in China and driven them down to small populations. Approximately one out of every four species of plants and animals in China is of concern regarding its conservation status: it may exist in relatively small populations, and is, therefore, relatively apt to become extinct. Coordination between different ministries in China that have different responsibilities, and coordination of the government itself with external agencies that are interested in these problems, has been less than perfect over the past years. A much higher level of coordinated effort will be necessary if the plants and animals of China are to be preserved for the future. A scheme for preserving the plants of the world is underway as part of the Convention on Biological Diversity of the United Nations, and hopefully, this will lead to the formation of adequate national programs for conserving plants, both in nature and in seed banks and botanical gardens, and that China will be an even more active participant in these efforts in the future.

The recent supportive efforts by the Chinese Academy of Sciences for herbaria and efforts for systematic botany throughout China are very important in building a database regarding the conservation status of all plant species. The efforts that are beginning now to strengthen botanical gardens in China for conservation, both those owned by the Academy and others, will be very important in achieving these objectives.

5. CHINESE MEDICINAL PLANTS AND TRADITIONAL CHINESE MEDICINE

The tradition of medicinal plants in China goes back nearly 3,000 years to the legendary Emperor Shennung, and has continued unbroken up to the present. In the previous chapter by professor Hui, there is a good overview of how the interactions between the Chinese people and medicinal plants not only have developed historically, but also exist now. Like the biodiversity of China itself, the medicinal plants, which may constitute one out of every three plants in China, constitute an extremely important resource for the future sustainable development of China, as well as an unparalleled world heritage, something in which people all over the world are deeply and legitimately interested. The conservation problems are even more serious for medicinal plants, because they are obviously gathered for use, both within China, and increasingly for supply to world markets. Only about 15 percent – about one-sixth, of the medicinal products produced from plants in China are produced from plants in cultivation, and that is a really serious problem for the future, as the demand increases in Europe, Japan, the United States, and China itself. There is a widespread belief that plants gathered in nature are more efficacious in traditional Chinese medicine than plants that are grown in cultivation. The kinds of scientific studies that Professor Hui discussed in the previous chapter will eventually enable us to understand what the active compounds are and how to increase the efficacy of cultivated medicinal plants, especially with the help of modern genetic technology, which is obviously the way we have to go in the future if there is going to be enough of a supply of the plants to meet the increasing demand.

In China, as in India, which is the other country with a really important tradition of medicinal plants, these are part of a holistic attitude about life and a different style of life from the style of life in the West. These indigenous systems of using plants really involve a whole philosophy and a whole way of living that differs greatly from certain Western industrialized countries. The ways to bridge the gap between these systems, to understand them, to bring them into focus and to move together into the world of the future are properly, and very importantly, matters to be considered here in working out a common future. One of the most interesting aspects of the whole system is the transition in Western medicine from curing diseases such as cancer, once they occur, to understanding the origins of those diseases and moving to preventative uses of plant products for more widespread use in diets so that these diseases do not occur in the first place.

The importance of the Human Genome Project has already been mentioned as an important element in beginning to use plant-derived compounds in these preventative approaches in a more sophisticated way. But another extremely important aspect of the situation that is already integrated into TCM is the variation among individual human beings, which is factored into TCM automatically as practitioners prescribe different combinations of drugs for individual people. In traditional western medicine, this has not been sorted out nearly as well, with its reliance on such things as minimum dietary allowances, and so forth, when, in fact, those allowances differ radically for elderly people, middle-aged people, teenagers, children, women, men, and all the rest. There are huge differences that increasingly will be taken into account in medicine in the future. And at an even more sophisticated level, the understanding of individual genetic variation will be able to be used in terms of finding individual treatments in relation to individual compounds.

6. CONCLUSION

To a certain extent, it is obvious in the pharmaceutical industry that pure combinatorial chemistry has run its course, and that it has delivered a lot of the value that it can, with the available molecules. Now, the pendulum has swung back again to consider seriously the marvelous array of secondary metabolites that plants use to protect themselves against insects, herbivores, and diseases, and that will lead us to a more sophisticated knowledge of plant chemistry. What can be accomplished in the future is extraordinarily beyond anything that one can even begin to imagine now.

The extraordinarily rich diversity of Chinese plants, which has been studied and nurtured so wonderfully by the Chinese people over some 3,000 years, comes into world focus as a genuine treasure for the Chinese people, and for people all over the world.

Chapter 3

NATURAL PRODUCTS DRUG DISCOVERY AND DEVELOPMENT AT THE UNITED STATES NATIONAL CANCER INSTITUTE

GORDON M. CRAGG* AND DAVID J. NEWMAN
*Natural Products Branch, Developmental Therapeutics Program, Division of Cancer
Treatment and Diagnosis, National Cancer Institute-Frederick, Fairview Center, Suite 206,
P. O. Box B, Frederick, Maryland 21702-1201.
Author to whom enquiries should be addressed

Abstract: Between 1960 and 1982, the National Cancer Institute (NCI) screened over 180,000 microbial-derived, some 16,000 marine organism-derived, and over 114,000 plant-derived extracts. A number of clinically effective chemotherapeutic agents were developed, mainly through collaborative efforts with the public and private sectors. These agents include paclitaxel, camptothecin derivatives, various anthracyclines, bleomycin, actinomycin and mitomycin. Since 1986, collections of plants and marine invertebrates have been performed in over 25 tropical and subtropical countries worldwide through contracts with botanical and marine biological organizations, working in close collaboration with source country organizations and subject to agreements with the source country authorities. Over 120,000 extracts are stored at low temperatures in the NCI Natural Products Repository and are made available to the scientific community for testing in screens related to all human diseases, subject to the signing of Material Transfer Agreements, which protects the rights of all parties, particularly those of the source countries. In addition, Memoranda of Understanding (MOU) have been signed with qualified organizations in over ten source countries for direct collaboration in the drug discovery and development process. A major goal of these collaborations is to promote drug discovery in the source country, with NCI collaborating in the preclinical and clinical development stages.

1. INTRODUCTION

The National Cancer Institute (NCI) was established in 1937, with a mission "to provide for, foster and aid in coordinating research related to cancer." In 1955, NCI set up the Cancer Chemotherapy National Service Center (CCNSC) to coordinate a national voluntary cooperative cancer chemotherapy program, involving the procurement of drugs, screening, pre-

Yuan Lin (ed.), Drug Discovery and Traditional Chinese
Medicine: Science, Regulatory and Globalization, 19-32.
©2001 *Kluwer Academic Publishers. Printed in the Netherlands.*

clinical studies, and clinical evaluation of new agents. By 1958, the initial service nature of the organization had evolved into a drug research and development program with input from academic sources and substantial participation of the pharmaceutical industry. The responsibility for drug discovery and pre-clinical development at NCI now rests with the Developmental Therapeutics Program (DTP), a major component of the Division of Cancer Treatment and Diagnosis (DCTD). Thus, for the past forty years, NCI has provided a resource for the pre-clinical screening of compounds and materials submitted by grantees, contractors, pharmaceutical and chemical companies, and other scientists and institutions, public and private, worldwide. The NCI has also played a major role in the discovery and development of many of the available commercial and investigational anticancer agents. During this period, more than 400,000 chemicals, both synthetic and natural, have been screened for antitumor activity.

Initially, most of the materials screened were pure compounds of synthetic origin, but the program also recognized that natural products were an excellent source of complex chemical structures with a wide variety of biological activities. From 1960 to 1982, over 180,000 microbial-derived, some 16,000 marine organism-derived, and over 114,000 plant-derived extracts were screened for antitumor activity, mainly by the NCI. As illustrated above, a number of clinically effective chemotherapeutic agents have been developed (Cragg et al., 1999).

2. ANTICANCER AGENTS FROM NATURAL SOURCES

2. 1 PLANT SOURCES

Plants have a long history of use in the treatment of cancer (Hartwell, 1982), though many of the claims for the efficacy of such treatment should be viewed with some skepticism because cancer, as a specific disease entity, is likely to be poorly defined in terms of folklore and traditional medicine (Cragg et al., 1994). Of the plant-derived anticancer drugs in clinical use, the best known is the so-called vinca alkaloids, vinblastine and vincristine, isolated from the Madagascar periwinkle, *Catharanthus roseus. C. roseus* was used by various cultures for the treatment of diabetes, and vinblastine and vincristine were first discovered during an investigation of the plant as a source of potential oral hypoglycemic agents. Therefore, their discovery may be indirectly attributed to the observation of an unrelated medicinal use of the source plant. The two clinically active agents, etoposide and teniposide, which are semisynthetic derivatives of the natural product epipodophyllotoxin, may be more closely linked to a plant originally used for the treatment of "cancer".

Epipodophyllotoxin is an isomer of podophyllotoxin that was isolated as the active antitumor agent from the roots of various species of the genus *Podophyllum*. These plants possess a long history of medicinal use by early American and Asian cultures, including the treatment of skin cancers and warts (Cragg et al., 1994).

More recent additions to the armamentarium of naturally derived chemotherapeutic agents include the taxanes and camptothecins. Paclitaxel initially was isolated from the bark of *Taxus brevifolia*, collected in Washington State, as part of a random collection program by the US Department of Agriculture for the NCI (Cragg, Schepartz, et al., 1993). Several Native American tribes have reported using various parts of *T. brevifolia* and other *Taxus* species (e.g., *canadensis, baccata*) for the treatment of some non-cancerous conditions (Hartwell, 1982), while the traditional Asiatic Indian (Ayurvedic) medicine system has reported using the leaves of *T. baccata* (Kapoor, 1990), with one reported use in the treatment of "cancer" (Hartwell, 1982). Paclitaxel, along with several key precursors (the baccatins), occurs in the leaves of various *Taxus* species, and the ready semi-synthetic conversion of the relatively abundant baccatins to paclitaxel, as well as active paclitaxel analogs, such as docetaxel (Cortes and Pazdur, 1995), has provided a major, renewable natural source of this important class of drugs. Likewise, the clinically active agents, topotecan (hycamptamine), irinotecan (CPT-11), 9-amino- and 9-nitro-camptothecin, are semi-synthetically derived from camptothecin, isolated from the Chinese ornamental tree, *Camptotheca acuminata* (Potmeisel and Pinedo, 1995). Camptothecin (as its sodium salt) was advanced to clinical trials by the NCI in the 1970s, but was later dropped because of its severe bladder toxicity.

Other examples of plant-derived anticancer agents currently in investigational use, are homoharringtonine, isolated from the Chinese tree, *Cephalotaxus harringtonia* var. *drupacea* (Sieb and Zucc.), and elliptinium, a derivative of ellipticine, isolated from species of several genera of the *Apocynaceae* family, including *Bleekeria vitensis*, a Fijian medicinal plant with reputed anticancer properties (Cragg et al., 1994). Homoharringtonine has shown efficacy against various leukemias, while elliptinium is marketed in France for the treatment of breast cancer (Cragg, Boyd, et al., 1993). The flavone, flavopiridol, currently in Phase I clinical trials, is scheduled to be advanced to Phase II trials against a broad range of tumors (Christian et al., 1997). While flavopiridol is totally synthetic, the basis for its novel structure is a natural product isolated from *Dysoxylum binectariferum* (Naik et al., 1988). Ipomeanol, a pneumotoxic furan derivative produced by sweet potatoes (*Ipomoea batatas*) infected with the fungus, *Fusarium solani*, has been in clinical trials for treatment of lung cancer (Cragg, Boyd, et al., 1993).

22

2.2 MICROBIAL SOURCES

Antitumor antibiotics are amongst the most important of the cancer chemotherapeutic agents, which include members of the anthracycline, bleomycin, actinomycin, mitomycin and aureolic acid families (Foye, 1995). Clinically useful agents from these families are the daunomycin-related agents, daunomycin itself, doxorubicin, idarubicin and epirubicin; the glycopeptidic bleomycins A_2 and B_2 (blenoxane); the peptolides exemplified by dactinomycin; the mitosanes such as mitomycin C; and the glycosylated anthracenone, mithramycin. All were isolated from various *Streptomyces* species. Other clinically active agents isolated from *Streptomyces* include streptozocin and deoxycoformycin.

Microbial metabolites in past or present clinical trials for the treatment of cancer include acivicin, aclacinomycin, deoxyspergualin, echinomycin, elsametrocin, fostriecin, menogaril, porfiromycin, quinocarmycin and rhizoxin, as well as the glycinate of aphidicolin. Microbial products predominate amongst the agents under development by the Division of Cancer Treatment and Diagnosis (DCTD) of the NCI (Foye, 1995). These include UCN-01 (7-hydroxystaurosporine), isolated from a *Streptomyces* species, and FR901228, a novel bicyclic depsipeptide isolated from a *Chromobacterium violaceum* strain, as well as derivatives of quinocarmycin (DX-52-1), spicamycin (KRN5500), CC-1065 (bizelesin), tetracycline (COL-3), rapamycin, and rebeccamycin. Recent exciting discoveries are the epothilones isolated from myxobacteria (Nicolaou, 1998). This class of compounds has been shown to act by a similar mechanism of action to paclitaxel and could complement the taxanes as chemotherapeutic agents.

The large number of microbial agents reflects the major role played by the pharmaceutical industry in this area of drug discovery and development. Generally, industry has focused on the *Actinomycetales,* but expansion of research efforts, often supported by government funding, to the study of organisms from diverse environments, such as shallow and deep marine ecosystems and deep terrestrial subsurface layers, has demonstrated their potential as a source of novel bioactive metabolites (Colwell, 1997).

2.3 MARINE SOURCES

The first notable discovery of biologically active compounds from marine sources was the serendipitous isolation of the C-nucleosides, spongouridine and spongothymidine, from the Caribbean sponge, *Cryptotheca crypta*, in the early 1950s. These compounds were found to possess antiviral activity, and synthetic analog studies eventually led to the development of cytosine arabinoside (Ara-C) as a clinically useful anticancer agent approximately 15 years later (McConnell et al., 1994), together with Ara-A as an anti-viral agent. The systematic investigation of marine environments, as sources of novel biologically active agents, only began in earnest in the mid-1970s. During the decade from 1977-1987, about 2,500 new metabolites were reported from a variety of marine organisms. These studies have clearly demonstrated that the marine environment is a rich source of bioactive compounds, many of which belong to totally novel chemical classes not found in terrestrial sources (Carte, 1996).

As yet, no compound isolated from a marine source has advanced to commercial use as a chemotherapeutic agent, though several are in various phases of clinical development as potential anticancer agents. The most prominent of these is bryostatin 1, isolated from the bryozoan, *Bugula neritina* (McConnell et al., 1994). This agent exerts a range of biological effects, thought to occur through modulation of protein kinase C, and has shown some promising activity against melanoma in Phase I studies (Philip et al., 1993). Phase II trials are either in progress or are planned against a variety of tumors, including ovarian carcinoma and NHL.

The first marine-derived compound to enter clinical trials was didemnin B, isolated from the tunicate, *Trididemnum solidum* (McConnell et al., 1994). Unfortunately, it has failed to show reproducible activity against a range of tumors in Phase II clinical trials, while always demonstrating significant toxicity. Ecteinascidin 743, a metabolite produced by another tunicate *Ecteinascidia turbinata*, has significant *in vivo* activity against the murine B16 melanoma and human MX-1 breast carcinoma models, and currently is scheduled for Phase II clinical trials in Europe and the United States (per. comm. G. Faircloth, PharmaMar). The sea hare, *Dolabella auricularia,* an herbivorous mollusk from the Indian Ocean, is the source of more than 15 cytotoxic cyclic and linear peptides, called dolastatins. The most active of these, the linear tetrapeptide, dolastatin 10, has been chemically synthesized and is currently in Phase I clinical trials (Carte, 1996).

Sponges are traditionally a rich source of bioactive compounds in a variety of pharmacological screens (Carte, 1996). In the cancer area, halichondrin B, a macrocyclic polyether initially isolated from the sponge, *Halichondria okada*, in 1985, was accepted for preclinical development by the NCI in 1992. Analogs derived from the total synthesis of halichondrin B have shown superior activity to the natural product (M.J.Towle et al., AACR, March 2000, Abstract 1370) and are now in advanced preclinical development by the NCI, in collaboration with Eisai Research Institute. Another sponge-derived agent of considerable interest is discodermolide, isolated from *Discodermia dissoluta*, which has been shown to act by a similar mechanism of action to paclitaxel (ter Haar et al., 1996). Discodermolide is currently in preclinical development by Novartis. Another active agent, eleutherobin, isolated from the soft coral, *Eleutherobia aurea*, also acts in a similar manner to paclitaxel (Long et al., 1998). In view of the clinical success of paclitaxel and related analogs, such as docetaxel, there is considerable interest in other classes of compounds sharing the same basic mechanism of action.

This interest in nature as a source of potential chemotherapeutic agents continues. An analysis of the number and sources of anticancer and anti-infective agents, reported mainly in the Annual Reports of Medicinal Chemistry from 1984 to 1995 covering the years 1983 to 1994, indicates that over 60% of the approved drugs developed in these disease areas are of natural origin (Cragg et al., 1997).

3. CURRENT STATUS OF THE NCI NATURAL PRODUCTS DRUG DISCOVERY AND DEVELOPMENT PROGRAM

3.1 CONTRACT COLLECTIONS

Since 1986, contracts for the cultivation and extraction of fungi and cyanobacteria and for the collection of marine invertebrates and terrestrial plants were initiated in 1986. With the exception of fungi and cyanobacteria, these programs continue to operate. Marine organism collections originally focused in the Caribbean and Australasian, but have now expanded to the Central and Southern Pacific and to the Indian Ocean (off East and Southern Africa) through a contract with the Coral Reef Research Foundation, which is based in Palau in Micronesia. Terrestrial plant collections have been carried out in over 25 countries in tropical and subtropical regions worldwide through contracts with the Missouri Botanical Garden (Africa and Madagascar), the New York Botanical Garden (Central and South America), and the University of Illinois at Chicago (Southeast Asia), and have been expanded to the continental United States through a contract with the Morton Arboretum.
In carrying out these collections, the NCI contractors work closely with qualified organizations in each of the source countries. Botanists and marine biologists from source country organizations collaborate in field collection activities and taxonomic identifications, and their knowledge of local species and conditions is indispensable to the success of the NCI collection operations. Source country organizations provide facilities for the preparation, packaging, and shipment of the samples to the NCI's Natural Products Repository (NPR) in Frederick, Maryland. The collaboration between the source country organizations and the NCI collection contractors, in turn, provides support for expanded research activities by source country biologists, and the deposition of a voucher specimen of each species collected in the national herbarium or repository is expanding source country holdings of their biota. When requested, NCI contractors also provide training opportunities for local personnel through conducting workshops and presentation of lectures. In addition, through its Letter of Collection (LOC) and agreements based upon it, the NCI invites scientists nominated by Source Country Organizations to visit its facilities, or equivalent facilities in other

approved U.S. organizations for 1-12 months to participate in collaborative natural products research. Representatives of most of the source countries have visited the NCI and contractor facilities for shorter periods to discuss collaboration (Mays et al., 1997). The LOC also dictates terms of benefit sharing and use of source country resources in the event of the licensing and development of a promising drug candidate. It should be noted that the formulation of the NCI policies for collaboration and compensation embodied in the Letter of Collection predated the drafting of the United Nations Convention on Biological Diversity in Rio de Janeiro by some four years. Contract collections of plants are now being de-emphasized in favor of establishing direct collaborations with qualified organizations in the source countries (discussed below).

Dried plant samples (0.3-1 kg dry weight) and frozen marine organism samples (~ 1 kg wet weight) are shipped to the NPR in Frederick where they are stored at -20°C prior to extraction with a 1:1 mixture of methanol: dichloromethane and water to give organic solvent and aqueous extracts. All extracts are assigned discreet NCI numbers and returned to the NPR for storage at -20°C until requested for screening or further investigation. After testing in the *in vitro* human cancer cell line screen, active extracts are subjected to bioassay-guided fractionation to isolate and characterize the pure, active constituents. Agents showing significant activity in the primary *in vitro* screens are selected for secondary testing in several *in vivo* systems. Those agents exhibiting significant *in vivo* activity are advanced into pre-clinical and clinical development.

3.2 DISTRIBUTION OF EXTRACTS FROM THE NCI NATURAL PRODUCTS REPOSITORY

In carrying out the collection and extraction of thousands of plant and marine organism samples worldwide, the NCI established the NPR, which is a unique and valuable resource for the discovery of potential new drugs and other bioactive agents. The rapid progress made in the elucidation of mechanisms underlying human diseases has resulted in a proliferation of molecular targets available for potential drug treatment. The adaptation of these targets to high throughput screening processes has greatly expanded the potential for drug discovery. In recognition of this potential, the NCI has developed policies for the distribution of extracts from the NPR to qualified organizations for testing in screens related to all human diseases, subject to the signing of a legally-binding Material Transfer Agreement (MTA), which protects the rights of all parties (see DTP Homepage at dtp.nci.nih.gov). One of the key terms of the MTA is the requirement that the recipient organization negotiate suitable terms of collaboration and compensation with the source country (ies) of any extract(s) that yield agents that are developed towards clinical trials and possible commercialization.

3.3 SCREENING AGREEMENT

In the case of organizations wishing to have pure compounds tested in the NCI drug screening program, such as pharmaceutical and chemical companies or academic research groups, the DTP/NCI has formulated a screening agreement that includes terms stipulating confidentiality, patent rights, routine and non-proprietary screening and testing versus non-routine, and levels of collaboration in the drug development process. Individual scientists and research organizations wishing to submit pure compounds for testing generally consider entering into this agreement with the NCI DCTD (available on the DTP website at http://dtp.nci.nih.gov.), and may also submit details of the compounds on-line through the DTP website. Should a compound show promising anticancer activity in the routine screening operations, the NCI would propose the establishment of a more formal collaboration, such as a Cooperative Research and Development Agreement (CRADA) or a Clinical Trial Agreement (CTA).

3.4 PRECLINICAL DRUG DEVELOPMENT

Those agents showing significant *in vivo* activity in appropriate animal models are evaluated by the DCTD Drug Development Group (DDG), and those meeting the DDG selection criteria are entered into preclinical development through the DDG process. The key steps involved in the preclinical development process are:

1) Development of an adequate supply of the agent to permit pre-clinical and clinical development.
2) Formulation studies to develop a suitable vehicle to solubilize the drug for administration to patients, generally by intravenous injection or infusion in the case of cancer.
3) Pharmacological evaluation to determine the best route and schedule of administration to achieve optimal activity of the drug in animal models, the half-lives and bio-availability of the drug in blood and plasma, the rates of clearance and the routes of excretion, and the identity and rates of formation of possible metabolites.
4) IND-directed toxicological studies to determine the type and degree of major toxicities in rodent and dog models. These studies help to establish the safe starting doses for administration to human patients in clinical trials.

Alternatively, agents may be developed through Rapid Access to Intervention Development (RAID), a program designed to facilitate translation to the clinic of novel, scientifically meritorious therapeutic interventions originating in the academic community. The RAID process makes available to the academic research community, on a competitive basis, NCI resources for pre-clinical development of a drug and functions as a collaboration between the NCI and the originating laboratory, with tasks apportioned to either the NCI or the originating laboratory, depending on the facilities and expertise available in the latter. While the RAID process is similar to the DDG process discussed above, the products of the RAID program are returned directly to the originating laboratory for proof-of-principle clinical trials. The RAID process cannot be used by private industry (which can interact with NCI through the DDG process), nor can it be used to develop a product already licensed to a company. However, the existence of research collaborations between the academic investigators and companies does not affect the eligibility for support from RAID for an individual product, provided the product is not licensed to a company.

3.5 CLINICAL DEVELOPMENT

Phase I studies are conducted to determine the maximum tolerated dose (MTD) of a drug in humans and to observe the sites and reversibilities of any toxic effects. In contrast to trials with agents directed at other diseases, all patients in Phase I cancer trials have some form of the disease. Once the MTD has been determined and the clinicians are satisfied that no insurmountable problems exist with toxicities, the drug advances to Phase II clinical trials. These trials generally are conducted to test the efficacy of the drug against a range of different cancer disease types. In those cancers where significant responses are observed, Phase III trials are conducted to compare the activity of the drug with that of the best chemotherapeutic agents currently available for the treatment of those cancers. In addition, the new drug may be tried in combination with other effective agents to determine if the efficacy of the combined regimen exceeds that of the individual drugs used alone.

Much of the NCI drug discovery and development effort has been, and continues to be, carried out through collaborations with academic institutions, research organizations and the pharmaceutical industry worldwide. Many of the naturally derived anticancer agents were developed through such efforts. The DTP/NCI thus complements the efforts of the pharmaceutical industry and other research organizations through taking positive leads, which industry might consider too uncertain to sponsor, and conducting the "high risk" research necessary to determine their potential utility as anticancer drugs. In promoting drug discovery and development, the DTP/NCI has formulated various mechanisms for establishing collaborations with research groups worldwide.

3.6 SOURCE COUNTRY COLLABORATION

3.6.1 DRUG DISCOVERY

As discussed earlier, the collections of plants and marine organisms have been carried out in over 25 countries through contracts with qualified botanical and marine biological organizations working in close collaboration with qualified source country organizations. The recognition of the value of the natural resources (plant, marine and microbial) being investigated by the NCI and the significant contributions being made by source country scientists in aiding the performance of the NCI collection programs have led the NCI to formulate its LOC, specifying policies aimed at facilitating collaboration with, and compensation of, countries participating in the drug discovery program (Mays et al., 1997).

With the increased awareness of genetically-rich source countries to the value of their natural resources and the confirmation of source country sovereign rights over these resources by the U. N. Convention of Biological Diversity, organizations involved in drug discovery and development are increasingly adopting policies of equitable collaboration and compensation in interacting with these countries (Baker et al., 1995). Particularly in the area of plant-related studies, source country scientists and governments are committed to performing more of the operations in country, as opposed to the export of raw materials. The NCI has recognized this fact for several years, and has negotiated Memoranda of Understanding (MOU; Appendix A) with a number of source country organizations suitably qualified to perform in-country processing. In considering the continuation of its plant-derived drug discovery program, the NCI has de-emphasized its contract collection projects in favor of expanding closer collaboration with qualified source country scientists and organizations. Through the MOUs, the NCI wishes to promote anticancer drug discovery in source countries, and may assist in the training of source country scientists and the establishment of small *in vitro* cancer cell line screens. The discovery of the active agents in-country will ensure that the intellectual property rights will reside with source country organization scientists, and applications for patents protecting those rights may be submitted by the source country organization. The NCI may collaborate with source country organizations in the further preclinical and clinical development of agents meeting the selection criteria of the NCI's DDG (vide infra). A number of other organizations and companies have implemented similar policies (Baker et al., 1995). Through this mechanism collaborations have been established with organizations in Australia, Bangladesh, Brazil, China, Costa Rica, Fiji, Iceland, Korea, Mexico, New Zealand, Nicaragua, Pakistan, Panama, and South Africa.

3.6.2 DRUG DEVELOPMENT: THE CALANOLIDES

In 1988, an organic extract of the leaves and twigs of the tree, *Calophyllum lanigerum*, collected in Sarawak, Malaysia in 1987, through the NCI contract with the University of Illinois at Chicago (UIC) in collaboration with the Sarawak Forestry Department, showed significant anti-HIV activity. Bioassay-guided fractionation of the extract yielded (+)-calanolide A as the main *in vitro* active agent (Kashman et al., 1992). Attempted recollections in 1991 failed to locate the original tree, and collections of other specimens of the same species gave only trace amounts of calanolide A. In 1992, a detailed survey of *C. lanigerum* and related species was undertaken by UIC and botanists of the Sarawak Forestry Department. As part of the survey, latex samples of *Calophyllum teysmanii* were collected and yielded extracts showing significant anti-HIV activity. The active constituent was found to be (-)-calanolide B which was isolated in yields of 20 to 30%. While (-)-calanolide B is slightly less active than (+)-calanolide A, it has the advantage of being readily available from the latex which is tapped in a sustainable manner by making small slash wounds in the bark of mature trees without causing any harm to the trees. A decision was made by the NCI/DNC to proceed with the pre-clinical development of both the calanolides, and, in June of 1994, an agreement based on the NCI Letter of Collection was signed between the Sarawak State Government and the NCI. Under the agreement a scientist from the University of Malaysia Sarawak was invited to visit the NCI laboratories in Frederick to participate in the further study of the compounds.

The NCI obtained patents on both calanolides, and, in 1995, an exclusive license for their development was awarded to Medichem Research, Inc., a small pharmaceutical company based near Chicago. Medichem Research had developed a synthesis of (+)-calanolide A (Flavin et al., 1996) under a Small Business Innovative Research (SBIR) grant from the NCI. The licensing agreement specified that Medichem Research negotiate an agreement with the Sarawak State Government. Meanwhile, by late 1995, the Sarawak State Forestry Department, UIC, and the NCI had collaborated in the collection of over 50 kg of latex of *C. teysmanii*, and kilogram quantities of (-)-calanolide B have been isolated for further development towards clinical trials. Medichem Research, in collaboration with the NCI through the signing of a Cooperative Research and Development Agreement (CRADA) by which NCI is contributing research knowledge and expertise, has advanced (+)-calanolide A through pre-clinical development, and was granted an INDA for clinical studies by the U. S. Food and Drug Administration (FDA). The Sarawak State Government and Medichem Research formed a joint venture company, Sarawak Medichem Pharmaceuticals Incorporated (SMP), in late 1996, and SMP has sponsored Phase I clinical studies with healthy volunteers. It has been shown that doses exceeding the expected levels required for efficacy against the virus are well tolerated. Trials using patients infected with HIV-1 are were initiated in early 1999.

The development of the calanolides is an excellent example of collaboration between a source country (Sarawak, Malaysia), a company (Medichem Research, Inc.) and the NCI in the development of promising drug candidates, and illustrates the effectiveness and strong commitment of the NCI to policies promoting the rights of source countries to fair and equitable collaboration and compensation in the drug discovery and development process. The development of the calanolides has been reviewed as a "Benefit-Sharing Case Study" for the Executive Secretary of the Convention on Biological Diversity by staff of the Royal Botanic Gardens, Kew (ten Kate and Wells, 1998).

3.7 DEVELOPMENTAL THERAPEUTICS PROGRAM WWW HOMEPAGE

The NCI DTP offers access to a considerable body of data and background information through its World Wide Web homepage: http://dtp.nci.nih.gov/ Publicly available data include results from the human tumor cell line screen and AIDS antiviral drug screen, the expression of molecular targets in cell lines, and 2D and 3D structural information. Background information is available on the drug screen and the behavior of "standard agents", NCI investigational drugs, analysis of screening data by COMPARE (Boyd and Paull, 1995), the AIDS antiviral drug screen, and the 3D database. Data and information are only available on so-called "open compounds" which are not subject to the terms of confidential submission.

In providing screening data on extracts, they are identified by code numbers only; details of the origin of the extracts, such as source organism taxonomy and location of collection, may only be obtained by individuals or organizations prepared to sign agreements binding them to terms of confidentiality and requirements regarding collaboration with, and compensation of, source countries. Such requirements are in line with the NCI commitments to the source countries through its LOC and the MTA.

4. CONCLUSION

Over the past 45 years, the NCI, through worldwide collaboration, has been involved in the development of many of the anticancer drugs currently in clinical use, many of them derived from natural product sources. Through MOUs, with suitably qualified source country organizations, the NCI may establish collaborative agreements for the promotion of the in-country discovery novel anticancer drugs from natural sources, and may assist in the preclinical and clinical development of those agents which meet the selection criteria of the NCI Drug Development Group.

REFERENCES

J. T. Baker, R. P. Borris, B. Carte, G. M. Cragg, M. P. Gupta, M. M. Iwu, D. R. Madulid, V. E. Tyler. Natural product drug discovery and development: new perspectives on international collaboration. *J Nat Prod.* **58**, 1325-1357 (1995).

M. R. Boyd, K. D. Paull: Some practical considerations and applications of the National Cancer Institute *in vitro* anticancer drug discovery screen. Drug Dev Res 34: 91-109, 1995.

B.K. Carte. Biomedical Potential of Marine Natural Products. *Bio-Science* **46**, 271-286 (1996).

M. C. Christian, J. M. Pluda, T.C. Ho, S.G. Arbuck, A. J Murgo, E. A. Sausville. Promising new agents under development by the Division of Cancer Treatment, Diagnosis, and Centers of the National Cancer Institute. *Sem. Oncol.* **24**, 219-240 (1997).

R.R. Colwell. Microbial diversity: the importance of exploration and conservation. *J. Ind. Microbiol. & Biotech.* **18**, 302-307 (1997).

J. E. Cortes, R. Pazdur. Docetaxel. *J. Clin. Oncol.* **13** 2643-2655 (1995).

G. M. Cragg, M. R. Boyd, J. H. Cardellina II, M. R. Grever, S. A. Schepartz, K.M. Snader, M. Suffness. Role of plants in the National Cancer Institute drug discovery and development program. In: Human Medicinal Agents from Plants. In *Human Medicinal Agents from Plants.* Am. Chem. Soc. Symposium Series (AD Kinghorn, MF Balandrin, eds.), Vol. 534, pp. 80-95. Amer. Chem. Soc., Washington, DC (1993).

G. M.Cragg, M. R Boyd, J. H. *Cardellina* II, D. J.Newman, K. M.Snader, T. G. McCloud. Ethnobotany and the Search for New Drugs. In *Ethnobotany and the Search for New Drugs.* Ciba Foundation Symposium, (DJ Chadwick, J Marsh, eds.), Vol. 185, pp. 178-196. Wiley & Sons, Chichester, U.K. (1994).

Cragg, G.M., Boyd, M.R., Khanna, R., Newman, D.J., and Sausville, E.A., Natural Products Drug Discovery and Development. The United States National Cancer Institute Role. In: Phytochemicals in Human Health Protection, Nutrition and Plant Defense, Romeo, (ed.), Kluwer Academic/Plenum Publishers, New York, pp. 1-29 (1999).

G. M.Cragg, D. J. Newman, K. M. Snader. Natural Products in Drug Discovery and Development. *J. Nat. Prod.* **60,** 52-60 (1997).

G. M. Cragg, S. A. Schepartz, M. Suffness, M. R. Grever. The taxol supply crisis. New NCI policies for handling the large-scale production of novel natural product anticancer and anti-HIV agents. *J. Nat. Prod.* **56**, 1657-1668, (1993).

M. T. Flavin, J. D. Rizzo, A. Khilevich, A. Kucherenko,. A. K. Sheinkman, V. Vilaychack,. L. Lin,. W. Chen, E.M. Greenwood, T. Pengsuparp,. J. Pezzuto, S.H. Hughes, T. M. Flavin,, M. Cibulski, W. A. Boulanger,. R.L. Shone,. Z.-Q. Xu. Synthesis, chromatographic resolution, and anti-human immunodeficiency virus activity of ($\pm$)-calanolide A and its enantiomers. *J. Med. Chem.* **39,** 1303-1313 (1996).

W. O. Foye. *Cancer Chemotherapeutic Agents,* ACS Professional Reference Book, Amer. Chem. Soc., Washington, D. C. (1995).

J. L. Hartwell. *Plants Used Against Cancer,* Quarterman, Lawrence, Massachusetts (1982).

L.D. Kapoor. *CRC Handbook of Ayurvedic Medicinal Plants*. CRC Press, Boca Raton, Florida (1990).

Y. Kashman, K. R. Gustafson, R. W. Fuller, J. H. Cardellina, II, J. B. McMahon, M. J. Currens, R. W. Buckheit, S. H Hughes, G. M. Cragg, M. R. Boyd. The calanolides, a novel HIV-inhibitory class of coumarin derivatives from the tropical rainforest tree, Calophyllum lanigerum. *J. Med. Chem.* **35**, 2735-2743 (1992).

B. H. Long, J. M. Carboni, A. J. Wasserman, L. A. Cornell, A. M. Casazza, P. R. Jensen, T. Lindel, W. Fenical, C. R. Fairchild. Eleutherobin, a novel cytotoxic agent that induces tubulin polymerization, is similar to paclitaxel. *Cancer Research* **58**, 1111-1115 (1998).

T. D. Mays, K. D. Mazan, G. M. Cragg, M. R. Boyd. "Triangular Privity" - a working paradigm for the equitable sharing of benefits from biodiversity research and development. In *Global Genetic Resources: Access, Ownership, and Intellectual Property Rights,* (KE Hoagland & AY Rossman, eds.), pp. 279-298. Association of Systematics Collections, , Washington, D. C. (1997).

O. McConnell, R. E. Longley, F. E. Koehn. The discovery of marine natural products with therapeutic potential. In *The Discovery of Natural Products with Therapeutic Potential* (V. P. Gullo, ed.), pp. 109-174. Butterworth-Heinemann, Boston, (1994).

R. G. Naik, S. L. Kattige, S. V. Bhat, B. Alreja, N. J. de Sousa, R. H. Rupp. An antiflammatory cum immunomodulatory piperidinylbenzopyranone from *Dysoxylum binectariferum*: Isolation, structure, and total synthesis. *Tetrahedron* **44**, 2081-2086 (1988).

K. C. Nicolaou, F. Roschangar, D. Vourloumis. Chemical Biology of the Epothilones. *Angew. Chem. Int.* **37**, 2014-2045 (1998).

P. A. Philip, D. Rea, P. Thavasu, J. Carmichel, N. S. A. Stuart, H. Rockett, D. C. Talbot, T. Ganesan, G. R. Pettit, F. Balkwill, A. L. Harris. Phase I study of bryostatin 1: assessment of interleukin 6 and tumor necrosis factor alpha induction in vivo. The Cancer Research Campaign Phase I Committee. *J. Natl. Cancer Inst.* **85**, 1812-1818 (1993).

M. Potmeisel, H. Pinedo. Camptothecins: *New Anticancer Agents*. CRC Press, Boca Raton, Florida (1995).

K. Ten Kate, A. Wells. Benefit-Sharing Case Study. The access and benefit-sharing policies of the United States National Cancer Institute: a comparative account of the discovery and development of the drugs Calanolide and Topotecan. Submission to the Executive Secretary of the Convention on Biological Diversity by the Royal Botanic Gardens, Kew (1998).

E. ter Haar, R. J. Kowalski, C. M. Lin, R. E. Longley, S. P. Gunasekera, H. S. Rosenkranz, B. W. Day. Discodermolide, a cytotoxic marine agent that stabilizes microtubules more potently than taxol. *Biochemistry* **35**, 243-250 (1996).

M. J. Towle, K. A. Salvato, J. Burrow, et al., Highly Potent in vitro and in vivo Anticancer Activities of Synthetic Macrocyclic Ketone Analogs of Halichondrin B. *Proceedings of the 91st. Annual Meeting of the American Society for Cancer Research*. San Francisco, 1370 (2000).

Chapter 4

THE GLOBAL IMPORTANCE OF PLANTS AS SOURCES OF MEDICINES AND THE FUTURE POTENTIAL OF CHINESE PLANTS

JAMES S. MILLER
Missouri Botanical Garden; P.O. Box 299; St. Louis, MO 63166-0299

Abstract: It has been estimated that more than 25% of prescription pharmaceuticals contain plant-derived ingredients yet only a small percentage of the plants in the world have been evaluated for potential pharmaceutical use. Increased efforts to survey plants as sources of new drugs in recent years have stepped up the pace of the discovery of new bioactive compounds from plants, and many programs will continue to contribute to this in the near future. With an estimated 30,000 species of plants, many with a long history of use as traditional medicines, China has played an important role in the development of presently used pharmaceuticals. Chinese plants also have promise to contribute useful medicines in the future. Several factors are contributing to the current interest in Chinese plants, particularly those used in Traditional Chinese Medicine. Recently programs have been initiated for clinical trials of herbal products from Chinese species. In addition, advances in pharmaceutical screening and evaluation of plants against a broader range of targets is increasing the potential for the discovery of new pharmaceutical and nutritional products from Chinese medicinal plants. Plants have always played an important role as a source of medicines both in a western sense and in a traditional sense. This review will cover the historical importance of plants in human health care and will examine three areas in which Chinese plants may be of particular importance in the future. New trends in natural products research have great importance for the potential use of Chinese plants for both the discovery of novel bioactive compounds and also the scientific validation of traditional Chinese medicine.

Yuan Lin (ed.), Drug Discovery and Traditional Chinese Medicine: Science, Regulatory and Globalization, 33-42.
©2001 *Kluwer Academic Publishers. Printed in the Netherlands.*

1. INTRODUCTION

While a general recognition of the importance of plants as a source of medicines exists, few people realize the full role that biological systems have played and the percentage of pharmaceuticals that owe part of their discovery and development to the natural world. The estimates for the percentage of pharmaceutical products that contain plant-derived ingredients range from about 25% (Farnsworth, 1977; Farnsworth et al., 1985) to more than 50% (Griffo et al., 1997), depending on how they are calculated. But these figures are low and do not take several factors into account. These estimates are based solely on plant-derived compounds and ignore other organisms that yield natural products. They underestimate natural products providing chemical scaffolds for the semi-synthetic development of new therapeutically useful compounds. They also discount the fact that a very large percentage of the world, including about 80% of people in developing countries, rely on herbal products as their primary source of healthcare. Plants are not the only organisms that yield natural products. Many anti-cancer drugs and other anti-infectives are derived from microbes, marine organisms, or other sources. Antibiotics were discovered from microbial sources and many continue to be derived from natural sources.

Natural products also serve as building blocks for drugs that are synthetically produced. Plants such as opium poppies, most widely known for yielding addictive compounds, provide us with models on which a number of opioid analygesics can be used that have great pharmaceutical potential (Psenak, 1998). When these two factors are taken into account, it is clear that the chemical constituents that make up the majority of western medicines come largely from natural sources. The 20-30% figures we hear for numbers of compounds that are derived directly from plants really underestimate the importance of natural sources in the development of western pharmaceuticals. Miller and Brewer (1992) were able to show that all of the top twenty selling drugs in the United States in 1988 had their discovery in some way related to natural products research.

These isolated bioactive compounds provide the basis for western pharmaceuticals, but access to western health care is extremely limited in many parts of the developing world. For example, it has been estimated that there is only one medical doctor for 70,000 people in rural Ghana, and that health services reach only 30% of the rural population (Abbiw, 1990). It is often estimated that more than 80% of the world's population are dependent on plants and herbal remedies for their primary source of health care. Unlike the global market in pure compounds, which form the basis of the pharmaceutical industry, the use of medicinal plants is generally a local industry, with plant species changing from one part of the world to another, depending on local availability.

2. EXPERIENCE OF THE MISSOURI BOTANICAL GARDEN

Researchers at the Missouri Botanical Garden (MBG) have been involved for approximately 15 years in formal programs for the development of new medicines with government, industrial, and university partners around the country. These programs span efforts to discover new pharmaceutical, agricultural, and nutritional products and make use of the extensive taxonomic ability of the researchers at the MBG in a variety of applied research areas.

MBG began a program collecting plants in Tropical Africa and Madagascar for the National Cancer Institute (NCI) in 1986 (Cragg et al., 1993). As part of that program, MBG botanists have collected about 16,000 plant samples from a wide variety of habitats in six countries. They are collected in extremely remote settings and shipped to the NCI facility in Frederick, Maryland. They have been processed, extracted, and then screened against 60 plus tumor cell lines and evaluated for their potential as anti-cancer drugs (Cragg et al., 1993).

This project has supported the collection and study of many interesting plants, some new to science, and has also yielded novel bioactive compounds that have been discovered and published by scientists at the NCI (e.g. Bernart et al., 1993; Bokesch et al., 1996; Boyd et al., 1994; Hallock et al., 1991). This has helped to promote our general understanding of the biological resources that are the raw material for these programs and bring attention to the conservation situation for these plants in the countries in which the MBG does work.

The MBG has also participated in one of the NIH-sponsored International Cooperative Biodiversity Groups (ICBG), a multi-institutional effort to combine natural products drug discovery with the aims of supporting economic development and conservation in participating developing countries (Rosenthal, 1997). The program initially focused on Suriname and later expanded to include Madagascar, has been coordinated through Virginia Polytechnic and State University (Kingston et al., 1999a; 1999b). The group has seven participating Associate Programs including: 1-Botany and Systematics (Missouri Botanical Garden), 2-Ethnobotany, Conservation, and Development (Conservation International), 3-Ethnobotany, Sample Processing & Phytomedicine Development (Centre National d'Applications et des Recherches Pharmaceutiques – Madagascar), 4-Sample Processing and Antimicrobial Drug Discovery (Bedrijf Geneesmiddelen Voorziening

Suriname), 5-Drug Discovery for Surinamese and Madagascan Plants (Bristol-Myers Squibb Pharmaceutical Research Institute), 6-Natural Products as Agrochemical Agents (Dow Agrosciences), and 7-Rain Forest Natural Products as Anticancer and Other Agents(Virginia Polytechnic and State University) (Rosenthal et al., 1999).

During the 1990s, the MBG also has participated in several bioprospecting programs with corporate partners. These programs have included discovery efforts spanning pharmaceutical, agricultural, and nutritional areas. In all of these programs, approximately 40,000 plant samples have been collected for evaluation.

This experience has also led to a consideration of the ethics and legal issues that affect international programs studying medicinal plants. The Convention on Biological Diversity entered into force in 1993; seven years after MBG botanists began collecting plants for the NCI. As part of that program, agreements had been signed with collaborating countries, in some cases, three or four years before the Convention entered into force. These original agreements from the NCI program are still widely held up as international models and were in full compliance with the convention (Mays et al., 1997). The convention mandates countries to regulate access to biological resources in exchange for an equitablc share of the benefits that arise from their use. Discussions of royalties and other direct financial benefits tend to dominate discussions of ethics in bioprospecting, but the benefits that can be derived from access to biological diversity can be tremendously diverse. The benefits that can arise from discovery programs are of three types: long-term, short-term, and public benefits.

3. BENEFITS OF BIOPROSPECTING

Long-term benefits include profit-sharing mechanisms, such as royalties or milestone payments for achieving certain developmental stages. These generally take a significant period of time before they arise and are dependent on successes in a program. Commitments that ensure that the source country is involved in the production of the raw biomass and continued research are also common and ultimately may be equally important in developing countries. The production of new drugs can have a tremendous positive impact on the economy in some of these countries, both from the development of high-value alternative crops and also from access to the technology that is necessary for research and production.

There are also short-term benefits, which materialize earlier in a program and depend on making rather uncommon discoveries. These include the collection of biological and biochemical data, opportunities for collaborative research, training and technology transfer, and direct contributions to increase institutional capacity, which is so critical in countries with high biological diversity and limited scientific research capacity. It is important to MBG that we can involve our collaborators in meaningful ways, share research opportunities, and use these as a real ethical basis for collaborative research. This includes appropriate co-authoring of publications, and exchange and re-patriation of the data that is collected, which ensures information will be available to the countries where it will have the most meaning. It provides opportunities for institutional capacity building that can be direct support for the improvement of the facilities, supplies or equipment in the institution all the way to helping develop the personnel that work there through a variety of training programs or opportunities to participate in sabbatical research programs. Once again, technology transfer can occur in either the long- or short-term.

There are also public benefits, such as the availability of new products that have health benefits, which improve the lives of humans in general. Public benefits also include the potential to boost the economies of the countries where plants are collected. All of this draws attention to the fact that plants are the basic resource upon which we depend as a source of new chemicals. Thus, bioprospecting programs help draw attention and promote conservation of these biological resources.

4. THE FUTURE IMPORTANCE OF CHINESE MEDICINAL PLANTS

Medicinal plant research is not a static field. It is evolving rapidly and three trends in new programs are particularly relevant to China. Recent changes in the study of medicinal plants expand the breadth of the genetic diversity studied, improve the rigor in the search for novel chemical structures, and increase the applications under consideration by evaluating whole plant remedies in a scientifically valid manner.

With an estimated 30,000 species, China is home to the richest temperate flora in the world, which is probably exceeded in total number of species only by Brazil, Columbia, and Indonesia. Associated with this unparalleled diversity is the richest history of use of plants as traditional medicines in the world. Perhaps one-third of the species in China have some history of use as medicines. Thus, the Chinese flora is particularly interesting for the discovery of novel bioactive compounds and also for the development and evaluation of more-traditional herbal plant remedies that may have a positive impact on human health. The Missouri Botanical Garden has been active in China for a number of years as one of the participating institutions in the Flora of China

Project. In collaboration with the California Academy of Sciences, Harvard University, the Institute of Botany, Academia Sinica, Beijing, the Jiangsu Institute of Botany, the Kunming Institute of Botany, the Natural History Museum, Paris, the Royal Botanic Garden, Edinburgh, the Royal Botanic Garden, Kew, the South China Institute of Botany, and the Smithsonian Institution, the project will catalog the estimated 30,000 species of plants in China, including notes on their use as traditional medicines.

The rich diversity of the Chinese flora, the strong history of traditional use of herbal medicines, and a resurgence of interest in screening for novel natural products, combined with an environment increasingly interested in the scientific validation of herbal remedies, ensure great interest in the plants of China. The following examples from MBG's own programs illustrate changes in the study of medicinal plants that represent great opportunities for the increased study of Chinese plants as new sources of medicines and indicate that they are likely to yield new products and useful discoveries.

Comprehensive sampling of the plant world involves broadening the kinds and types of plant material screened and kinds and types of medicinal targets against which the plants are evaluated. There are thousands of infectious diseases. In addition, the Human Genome Project recognizes thousands of physiological, genetically-based disorders, and the National Cancer Institute recognizes several hundred different kinds of cancers. The number of ailments that plague humans is essentially limitless. Each illness represents an area where a single plant species might prove efficacious. Thus, the number of pharmaceutical targets with which we can evaluate plant species is increasing tremendously. At the same time, we are also moving beyond the capacity to treat disease in a western sense and, through nutrition, trying to prevent disease or trigger changes in our immune systems or other parts of our physiology that prevent or slow the development disease.

Screening and evaluating plant diversity in a comprehensive way will involve a thorough survey of the genetic diversity in plants, including both wild plants and also the genetic diversity of cultivated plants that has developed during periods of domestication. One new cooperative program will be collecting plants with a history of human consumption that are generally regarded as safe. This will include sampling plants used as foods, flavorings, beverages, food additives, or herbal medicines. Many of these species have long histories of genetic diversification during periods of domestication and crop development. With perhaps the largest number of plant species used as herbal medicines and tremendous diversity of food plants, China is uniquely positioned with great genetic diversity of both wild and cultivated plants and will certainly play a major role as efforts to broaden the genetic diversity of plants screened is increased.

Most natural products discovery programs have involved collecting plants, producing extracts, and evaluating them in a variety of bioassays. The extracts may have been cleaned by removal of certain fractions, but crude extracts of plants present three problems in bioassays. For instance, these extracts are very complex mixtures of at least dozens, if not hundreds of individual chemical compounds. With so many compounds present at once, the possibility for interference between compounds, where one compound prevents the bioactivity of another, exists. In addition, there is also the possibility that activity of one compound will be masked by an opposing activity of a second compound in a complex mixture. Furthermore, within these complex mixtures, there are usually only a small number of compounds that are present at a high-enough concentration to register in a bioassay, with most of the components present at levels below where their potentially useful activity may be detected.

A new joint venture with a southern California company, Sequoia Sciences, will address these problems by increasing the rigor in the process to evaluate plant extracts. The MBG will be collecting and supplying plant material, and Sequoia Sciences will be processing and extracting the plants. Each extract will be fractionated repeatedly to build libraries of individual compounds or individual groups of compounds with very similar chromatographic profiles. This will allow evaluation of extracts in a much more thorough way than in past programs and will also facilitate rapid acquisition of analytical data required for structure determination of each compound.

The benefits of this process derive from evaluation compounds individually, rather than hundreds at a time. This will reduce the possibility for interference of activity between compounds and the likelihood for masking of activity that is found in complex crude extracts. It will also allow adjustment of the concentration of each compound to ensure an adequate amount for detection by bioassay. This should eliminate the situation where many of the chemical constituents of a plant are simply below the concentration that is required to register activity in an assay and are therefore not discovered. In addition, a variety of analytical data will be associated with each compound in the library to allow very rapid assignment to a class of compounds, help promote efficient de-replication, isolation, and identification of the active compounds, and effectively prioritize structures for further study and development. The chemical diversity of plants can most effectively be surveyed with this type of scientifically rigorous approach, and the rich biodiversity of China promises to yield novel chemical structures.

A final example is the evaluation of plants as herbal products, rather than as isolated individual compounds. A common explanation for the inability to isolate the active principals from a number of the plant species that are important in the herbal products industry is that activity results from the synergistic reaction of a mixture of compounds. Fractionation separates the necessary components and activity is lost. Yet efforts to decipher the complex mechanisms of action of these remedies have generally not been successful. The US National Institutes of Health have recently instituted, through the Office of Dietary Supplements, a new program that establishes national centers for the study of botanical dietary supplements. This program will attempt to isolate and identify active principals from botanical dietary supplements, understand mechanisms of action, document toxicity, evaluate clinical efficacy, and develop methods to ensure that ingredients used in manufacture are correctly identified. Attempts to collect clinical data about the efficacy of these products are particularly encouraging, as activity from multiple active ingredients has long complicated efforts to isolate active principles.

This effort has particular relevance to Traditional Chinese Medicine because it does not necessarily rely on isolation, purification, and identification of a single active component. It does however have the capacity to document that there is medical utility or positive human health benefit from something without having to actually identify and purify the chemical constituents. No country has a population more dependent on herbal medicines than China, but interest continues to increase in Europe and the United States as well and Chinese plants seem likely to be used in new parts of the world.

5. CONCLUSION

With its wealth of biological diversity and history of reliance on herbal medicine, China seems certain to play an important role in the future of medicinal plant research. Programs that survey the full range of genetic diversity of both wild and cultivated plants will broaden the base of plant material and increase likelihood of discoveries. Surveying plant extracts in a more rigorous, precise manner, so they are evaluated in a manner that ensures that all chemical constituents are assayed, should improve discovery rates. Adding a mechanism that allows evaluation of plants for their full range of activity rather than just surveys for pure chemical constituents should promote natural products drug discovery in general. The rate of discovery from these developments should improve, arenas in human health that will be affected will increase, and Chinese plants should play a major role.

REFERENCES

Bernart, M. W., Y. Kashman, M. Tischler, J. H. Cardellina II, and M. R. Boyd. 1993. Bershacolone, an unprecedented diterpene cyclobutene from *Maprounea Africana*. Tetrahedron Letters 34: 4461-4464.

Bokesch, H. R., T. C. McKee, M. J. Currens, R. J. Gulakowski, J. B. McMahon, J. H. Cardellina II, and M. R. Boyd. 1996. HIV-inhibitory gallotannins from *Lepidobotrys staudtii* Natural Product Letters 8: 133-136.

Boyd, M. R., Y. F. Hallock, J. H. Cardellina II, K. P. Manfredi, J. W. Blunt, J. B. McMahon, R. W. Buckheit, Jr., G. Bringmann, M. Schäffer, G. M. Cragg, D. W. Thomas, and J. G. Jato. 1994. Anti-HIV michellamines from *Ancistrocladus korupensis*. J. Med. Chem. 37: 1740-1745.

Cragg, G. M., M. R. Boyd, J. H. Cardellina II, M. R. Grever, S. A. Schepartz, K. M. Snader, and M. Suffness. 1993. Role of plants in the National Cancer Insitute drug discovery and development program. Pp. 80-95. *In*: A. D. Kinghorn and M. F. Balandrin (eds.) Human Medicinal Agents from Plants. American Chemical Society, Washington, D.C.

Farnsworth, N. R. 1977. Problems and prospects of discovering new drugs from higher plants by pharmacological screening. Pp. 1-22. *In*: H. Wagner and P. Wolf (eds.) New natural products with pharmacological, biological, or therapeutic activity. Springer-Verlag, New York.

Farnsworth, N. R., O. Akerele, A. Bingel, D.D. Soejarto, and Z. Guo. 1985. Medicinal plants in therapy. Bulletin of the World Health Organization 63: 965-981.

Grifo, F, D. Newman, A. S. Fairfield, B. Bhattacharya, and J. T. Grupenhoff. 1997. The origins of prescription drugs. Pp. 131-163. *In*: F. Grifo and J. Rosenthal (eds.) Biodiversity and Human Health. Island Press, Washington, D.C.

Hallock, Y. F., K. P. Manfredi, J. W. Blunt, J. H. Cardellina II, M. Schäffer, G. Bringmann, A. Y. Lee, J. Clardy, G. Francois, and M. R. Boyd. 1991. Korupensamines A-D, novel antimalarial alkaloids from *Ancistrocladus korupensis*. J. Org. Chem. 59: 6349-6355.

Kingston, D. G. I., M Abdel-Kader, B.-N. Zhou, S.-W. Yang, J. M. Berger, H. van der Werff, R. Evans, R. Mittermeier, S. Malone, L. Famalare, M. Guerin-McManus, J. H. Wisse, and J. S. Miller. 1999a. Biodiversity conservation, economic development, and drug discovery in Suriname. Pp. 39-59. *In*: S. J. and H. G. Cutler (eds.) Biologically Active Natural Products: Pharmaceuticals. CRC Press, Boca Raton.

Kingston, D. G. I, M. Abdel-Kader, B.-N. Zhoud, S.-W. Yang, J. M. Berger, H. van der Werff, J. S. Miller, R. Evans, R. Mittermeier, L. Famolare, M. Guerin-McManus, S. Malone, R. Nelson, E. Moniz, J. H. Wisse, D. M. Vyas, J. J. K. Wright, and S. Aboikonie. 1999b. The Suriname International Cooperative Biodiversity Group Program: lessons from the first five years. Pharmaceutical Biology 37 (Suppl.): 22-34.

Mays, T. D., K. Duffy-Mazan, G. Cragg, and M. Boyd. 1997. A paradigm for the equitable sharing of benefits resulting from biodiversity research and development. Pp. 267-280. *In*: F. Grifo and J. Rosenthal (eds.) Biodiversity and Human Health. Island Press, Washington, D.C.

Miller, J. S. and S. J. Brewer. 1992. The discovery of medicines and forest conservation. Pp. 119-134. *In*: R. P. and J. E. Adams (eds.) Conservation of Plant Genes Through DNA Banking and *in vitro* Biotechnology. Academic Press, San Diego.

Psenak, M. 1998. Biosynthesis of morphinane alkaloids. Pp. 159-188. *In*: J. Bernath (ed.) Poppy, the Genus Papaver. Harwood Academic Publishers, Amsterdam.

Rosenthal, J. 1997. Integrating drug discovery, biodiversity conservation, and economic development: early lessons from the International Cooperative Biodiversity Groups. Pp. 281-301. *In*: F. Grifo and J. Rosenthal (eds.) Biodiversity and Human Health. Island Press, Washington, D.C.

Rosenthal, J., D. Beck, A. Bhat, J. Biswas, L. Brady, K. Bridbord, S. Collins, G. Cragg, J. Edwards, A. Fairfield, M. Gottlieb, L.A. Gschwind, Y. Hallock, R. Hawks, R. Hegyeli, G. Johnson, G. Keusch, E. E. Lyons, R. Miller, J. Rodman, J. Roskoski, and D. Siegel-Causey. 1999. Combining high-risk science with ambitious social and economic goals. Pharmaceutical Biology 37 (Suppl.): 6-21.

Chapter 5

FOOD, MEDICINAL PLANTS, AND OTHER EDIBLE MATERIALS AS SOURCES OF BIOACTIVE COMPOUNDS THAT ENHANCE METABOLIC FITNESS AND IMPROVE HEALTH

STEVEN C. BOBZIN[1] AND PAUL BURN
Nutrition and Consumer Products; Monsanto Company, Pharmacia Corporation
800 North Lindbergh Boulevard
St. Louis, MO 63167
[1]*Corresponding author*

Abstract: Molecular nutrition is becoming an increasingly important scientific discipline of human nutrition. The systematic dissection of food and edible materials into their molecular components, followed by studying individual molecules in animal model systems and humans to investigate their nutritional and health benefits, is evolving rapidly. The identification, production, and marketing of these food-derived bioactives will not only have a significant impact on the development of new regimens for disease treatments, but will lead the way into a new era of preventing or postponing the onset of severe chronic diseases by maintaining optimal metabolic fitness and health. Sound scientific principles, rigorous clinical trials, and science-based regulatory processes will facilitate and help to define a new space in diet and dietary compound-based management of health and disease conditions.

Yuan Lin (ed.), Drug Discovery and Traditional Chinese Medicine: Science, Regulatory and Globalization, 43-54.
©2001 Kluwer Academic Publishers. Printed in the Netherlands.

1. UNTAPPED NICHE IN THE OVER THE COUNTER HEALTHCARE MARKET

Small start-up companies and divisions of large multinational pharmaceutical and food corporations recently have been launching research and development efforts into the field of human nutrition. The investment of significant financial and scientific resources by these organizations has been motivated by marketing and consumer research that indicated the existence of a sizable open and untapped niche in the over the counter (OTC) healthcare market. This untapped market niche is characterized by a consumer need for more self-managed preventative healthcare products, a healthcare practitioner desire for new treatments for patients with borderline disease conditions, and a retail belief that there is significant untapped consumer demand for self-directed products that have scientifically validated efficacy and safety for preventative healthcare. It is believed that this new area is distinct from current food, pharmaceutical (R_x), and OTC healthcare markets, as well as vitamin, mineral, and supplement (VMS) markets (Figure 1). The marketing strategy for these products will include advertising directly to the consumer and to the healthcare practitioner, using the limited labeling claims allowed under the *Dietary Supplement Health and Education Act* of 1994 (*DSHEA*). These limited labeling claims will be augmented by evidence for safety and efficacy provided by a science-driven approach to human clinical trials, and by documented identification and standardization of the active ingredient(s) in these products.

The business strategies created by many of the organizations seeking to establish themselves in this new area focus on these unmet market needs. These organizations aim to bring to the consumer products that *prevent or postpone the onset of disease states or treat people with borderline disease conditions*. Prevalent chronic diseases, such as hypertension, hypercholesterolemia, obesity, type II diabetes, osteoporosis, and arthritis are of particular interest due to their debilitating effects on an aging population. These diseases all possess various, characteristic pre-disease stages that correlate with and are characterized by a steady increase of well-known and accepted biomarkers that indicate the metabolic fitness and health of an individual. Biomarkers, such as blood pressure for hypertension, cholesterol level for hypercholesterolemia, body mass index for obesity, and glucose level for diabetes are very reliable disease stage indicators, which, in many cases, can be easily measured and monitored by the consumers themselves. The ability to self-monitor these biomarkers will allow consumers to follow their own progress and provide much needed feedback on the success of the product containing food-derived bioactive(s) in postponing or maintaining the desired health benefit. Recognition of these improvements will assure satisfactory compliance to dosing schedules and consumer loyalty to a product.

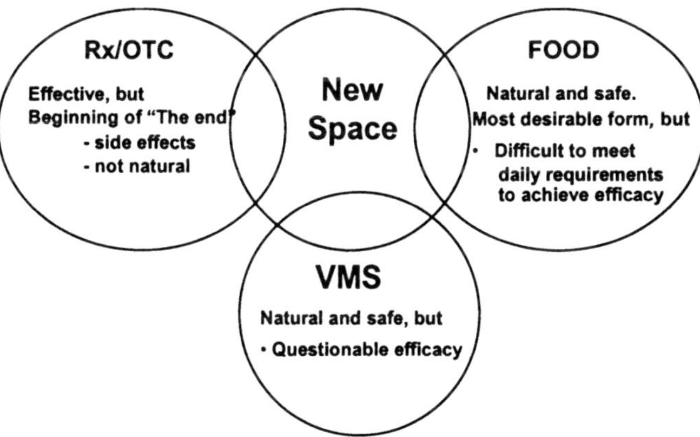

Figure 1. Diagram depicting the new market space where human nutrition is distinct from food, pharmaceutical (Rx) and over the counter (OTC) healthcare products, and vitamin, mineral, and supplements (VMS) markets.

Managing cardiovascular health by increasing metabolic fitness, using products of dietary origin, appears to be a particularly attractive entry into the marketplace for new businesses. Vascular diseases of the cardio and the cerebral system in humans account for nearly 55% of all deaths in developed countries, detract from the quality of life, and are responsible for the greatest percentage of healthcare costs of any disease category. Cardiovascular disease states can be monitored by two easily measured biomarkers, cholesterol and blood pressure. For example, serum cholesterol levels are a good indicator and a widely accepted biomarker of arterial health (Kritchevskey, 1995). In addition, mean serum cholesterol levels increase in the average population by age, in particular in postmenopausal women (Figure 2). Similarly, blood pressure is an accepted risk indicator of myocardial infarction (MI), stroke, and other acute vascular events. In several independent clinical studies, small reductions in blood pressure have been demonstrated to be directly correlated with a significant reduction in the risks for MI and stroke.

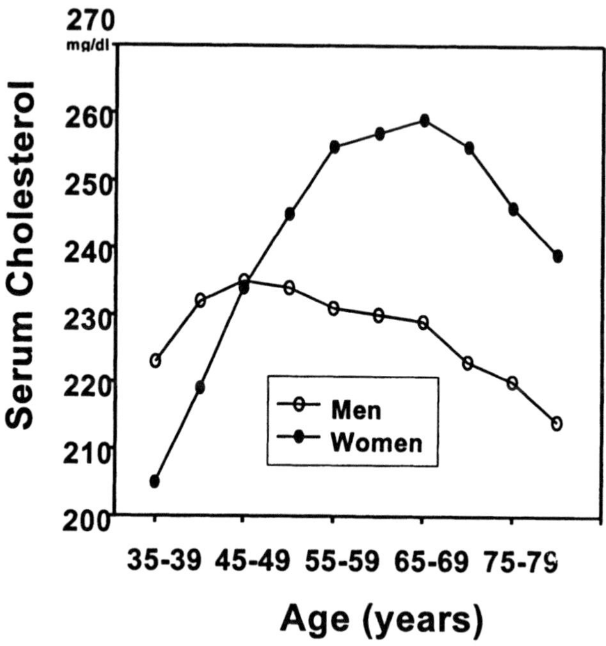

Figure 2. Trends in the mean serum cholesterol levels of men and women.

The product concept that has been generated from this market and consumer research is to create products containing compounds the desired health benefits from materials that are consumed by humans worldwide. Food plants, medicinal plants, and other edible materials consumed by humans in different parts of the globe may be the source of these bioactive constituents. The discovery of these health-enhancing components of food will provide a new category of condition-specific products targeted at maintaining or improving metabolic fitness (*i.e.* maintaining optimal blood pressure and serum cholesterol levels, or reducing them to healthy levels). These products will likely be presented as food supplements or dietary supplements in pill form. They will contain dietary-based molecules that will affect metabolic fitness and thus have a long-term impact on "condition-specific" health maintenance or disease prevention. A reduction in the progression rate of a disease or the maintenance of a healthy state will then be followed by measuring and monitoring the known biomarkers of the corresponding health condition. These products may contain purified active compound(s), crude extracts, or partially purified extracts. In any case, the bioactive component(s) will be determined, carefully quantified, and studied in animal models and humans to evaluate safety and efficacy of the marketed product.

2. RESEARCHING THE BIOACTIVE, MOLECULAR CONSTITUENTS OF FOOD

Human nutrition is more than a hundred year old science - this science has led to the discovery of many essential vitamins, minerals, amino acids, and fatty acids required for normal metabolic functioning of the human body. However, food contains nutrients that are not only capable of meeting caloric and general metabolic needs but that can also affect specific metabolic functions which can either initiate or retard the process of degeneration of the cells and tissues of the human body. Thus, identification of these bioactive, molecular constituents of food will allow researchers to study their intrinsic health benefits and to produce and provide them to people in pill form or as food supplements. In addition, these studies may provide future opportunities to apply modern biotechnological approaches to enhance the benefits of food by either including or excluding these bioactives from common food sources. A number of terms have been coined to refer to these classes of molecules - *"nutraceuticals"* is a commonly used name although other terms, such as *phytoceuticals, phytonutrients,* and *phytomedicines* have also been used (Tyler, 1999). *Nutraceuticals* and *functional foods* refer to products containing either bioactive ingredients or whole food products that positively affect physiological functions of the human body.

In recognition of this enormous opportunity, a large number of academic, government, and industry-based laboratories are now conducting research in the fields of natural products chemistry, pharmacology, and human clinical trials in an effort to identify new, safe, and efficacious components in food, medicinal plants, and other edible materials. In addition, many of these organizations and others, are engaged in leveraging the tools of molecular biology, genetics, genomics, and biotechnology to address issues of production and presentation of these bioactive components in food and feed (DellaPenna, 1999; Mazur *et al.,* 1999; Moffat, 1999). These rapidly evolving technologies can be linked to the science of human nutrition in a field that we will refer to as *"molecular nutrition"* (Burn and Kishore, 2000).

Research efforts are underway in many laboratories, both industrial and academic, throughout the world in an effort to isolate, identify, and establish the safety and efficacy of some of these as yet unrecognized bioactives. Researchers are investigating common food plants, medicinal plants, and

48

other edible materials collected from around the globe for components that may possess desired health enhancing effects. These laboratories are engaging in a true *discovery research* effort that will identify novel bioactive molecules with structures that have never before been seen. In addition, this effort is also likely to identify known structures and reveal previously unrecognized structure-function relationships for these known compounds.

3. PHARMACEUTICAL DISCOVERY RESEARCH APPROACH

In some of the laboratories conducting this type of work, the research effort has been organized using a pharmaceutical discovery research model. A disciplined pharmaceutical discovery research approach is believed to provide the fastest way to achieve results in identifying bioactives from food plants, medicinal plants, and edible material extracts. This approach includes the implementation of high throughput screening (HTS) technologies, such as automation and robotics, to allow rapid testing of thousands of natural extracts. These extracts are tested in assays formatted in 96-well microtiter plates against molecular targets, such as enzymes and receptors, and in functional assays using cell-based and reporter assays, which are believed to indicate the potential for molecule that provide a desired health benefit in humans. Once the HTS process identifies active extracts, natural products chemists must isolate and identify the active component(s) in each food plant extract. The knowledge of the active component(s) at this early stage will allow the characterization and standardization of materials for further testing. Teams of biologists and pharmacologists will then test these newly discovered bioactives *in vitro* and *in vivo* model systems to further establish safety, efficacy, and dosage. Finally, carefully controlled human clinical trials will be conducted in a randomized, double-blind, placebo-controlled manner using appropriate test group populations to establish safety and efficacy of these products in humans. The data for safety and efficacy, obtained from these clinical studies, will be far in excess of the current requirements needed to register these products with the US Food and Drug Administration (FDA) under the provisions of the *Dietary Supplement Health and Education Act* (DSHEA). Moreover, these data will provide additional information for the marketing and advertising of these products once they gain regulatory approval.

The success of any molecular discovery research effort is critically dependent upon the quality of the materials (chemicals or natural extracts) that go into the primary testing scheme. A rich and diverse collection of extracts from food plants, medicinal plants, and edible materials consumed by people throughout the world is essential to maximize the likelihood of identifying novel bioactives and previously unrecognized benefits of known components. Thus, a "Food Library" containing a collection of extracts of materials consumed worldwide must be assembled to drive research efforts in molecular nutrition. A Food Library that includes extracts of foods, spices, herbs, medicinal plants, meats, dairy products, and other materials consumed by humans will provide a unique screening resource. Furthermore, it will provide a source for proprietary discoveries of novel bioactives and previously unrecognized benefits of known components, which should allow regulatory approval of new products under DSHEA.

The strategy to focus on plants and materials consumed by humans worldwide is dictated by the United States FDA regulations, which allow regulatory approval under DSHEA. DSHEA, passed by Congress in 1994, states, *"The provisions of DSHEA define dietary supplements and dietary ingredients; establish a new framework for assuring safety; outline guidelines for literature, ...claims, ...labeling; and ...good manufacturing practices (GMP) regulations."* This document states that in order to approve a new product under the provisions of DSHEA, *"Supplements may contain new dietary ingredients...only if those ingredients have been present in the food supply as an article used for food in a form in which the food has not been chemically altered or there is a history of use, or some other evidence of safety exists that establishes that there is a reasonable expectation of safety when the product is used according to recommended conditions of use."* There are key questions which arise when one attempts to quantify historical usage by humans and provide evidence for a "reasonable expectation of safety". Answers to the following questions must be considered: How long have people consumed this? How frequent is consumption by an individual? How many people have consumed this? What is the common dietary dosage? In order to claim safety, the new product must have a daily dosage that's not significantly greater than the historical common dietary dosage determined from this information.

4. CONSIDERATIONS FOR CREATING A FOOD LIBRARY

All of these issues must be taken into serious consideration when defining the strategy for creating a Food Library. The acquisition of source materials for the Food Library should be prioritized by the extent of documented human consumption of the respective food source (Figure 3). Therefore, commonly consumed "staple" foods such as cereals, grains, fruits, and vegetables may be prioritized higher, and acquired before, less commonly consumed foods such as herbs, spices, and flavorings. Regionally consumed ethnic and cultural foods, and herbals and medicinal plants should be prioritized lower because of the relatively small population base consuming ethnic and cultural foods and the relatively low dietary dosage of herbals and medicinal plants consumed. A second consideration should be to maximize the biological diversity of the Food Library. It is commonly accepted in the field of natural products that biological diversity will yield the chemical diversity required to drive a successful molecular-based discovery program. An effort must be made to maximize the acquisition of samples from geographically (and culturally) diverse areas with attention to acquiring a biologically (taxonomically) diverse collection, thus providing the chemical diversity desired.

A Food Library may be compiled in many ways. It may be reasonable for a knowledgeable scientist from any research organization to embark on an effort to create such a library independently. This collection could focus on fruits, vegetable, grains, meats, dairy products, and other edible materials found at the local grocery store. This initial collection could then be expanded by collections at local farmers markets and visiting local ethnic markets. This approach would provide a starting point for a comprehensive Food Library but would be lacking in breadth of materials sampled and depth of information obtained about each sample. Establishing collaborations with academic, public, and government organizations such as universities, botanical gardens, and agricultural organizations would recruit the assistance of scientists with a much better understanding of the materials being obtained for the Food Library. The participation of these specialists would provide access to a greater diversity of materials to be incorporated into the Food Library and would provide better information about the samples collected. For instance, common names for food plants may vary in different areas, causing confusion as to the true identity of a plant. A trained botanist from a university or botanical garden may be able to provide the taxonomic identification for each sample collected. This would provide the Latin binomial (genus, species) that definitively identifies the plant in question.

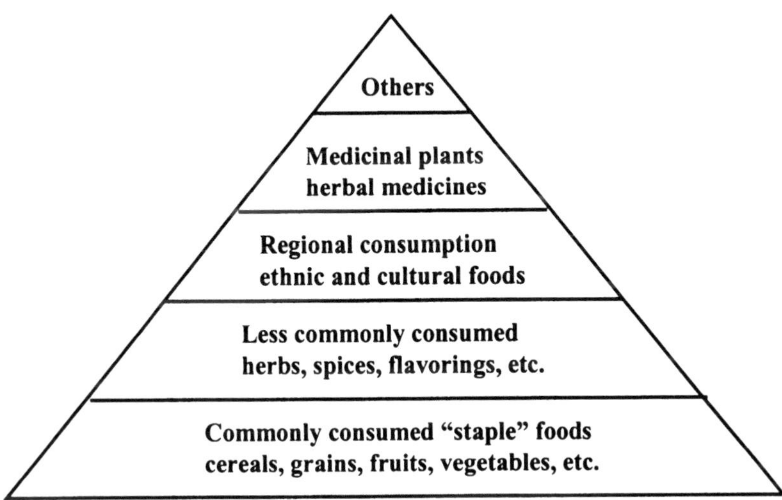

Figure 3. The Food Library Pyramid illustrating the potential prioritization of plant samples to be acquired for generating a Food Library and emphasizing the foundation of staple foods upon which a Food Library may be based.

5. ISSUES OF PROTECTING BIOLOGICAL DIVERSITY

Each of the food samples collected must be processed to produce extracts that are compatible with assays that are commonly used in the high throughput screening (HTS) process. Considering the effort that goes into obtaining raw materials, it is most desirable to release the full chemical diversity that exists in each of the natural materials collected for the Food Library. To achieve this goal, multiple extraction processes are likely to be used. This may include an extraction process using dichloromethane as a solvent to produce a lipophilic extract (containing fatty acids, sterols, etc.) and an ethanol extraction that produces an extract of moderate polarity (containing flavanoids, alkaloids, etc.). In addition, an aqueous extract may be prepared from a separate sample of material to provide a water-soluble extract (containing proteins, sugars, etc.). These extracts should be presented to the HTS assay in compatible solvents [dimethylsulfoxide (DMSO) for organic soluble extracts and water for aqueous extracts]. Another point to consider in the preparation of extracts for the HTS process is the standardization of the concentrations of crude extracts. If all extracts are prepared at a standard concentration, then assay results may be compared amongst different extracts allowing the prioritization of the extracts for further study.

6. ISSUES OF PROTECTING BIOLOGICAL DIVERSITY

All collections for a Food Library must be made with consideration of the United Nations Convention on Biological Diversity (CBD). The CBD was convened in Rio de Janeiro, Brazil by the United Nations in 1992 to address issues of protecting biodiversity worldwide and protecting the rights of sovereign nations and indigenous peoples. To date, over 150 countries have signed this treaty, but the US is not one of them. Nonetheless, organizations conducting this type of research should proceed in the spirit of the CBD, supporting the equitable sharing of benefits from the use of a country's biological resources and insuring that agreements are in place with the appropriate source country organization(s) before obtaining plant and natural materials. This is the only reasonable and ethical way to proceed.

7. THE INCORPORATION OF TRADITIONAL CHINESE MEDICINES

Traditional Chinese Medicines (TCMs) may be an important component for any Food Library. The plants and herbs used in TCMs may provide a useful addition to the collection, especially where the knowledge of traditional use is related to the disease states being studied. Many TCMs consist of a mixture of multiple plant species, each having components that contribute to the reported effect in humans. This observation may discourage one to investigate extracts of individual species from TCMs in a screening program, such as described earlier. However, in many cases, it has been shown that single species in a TCM mixture contain independently active components and other species in the TCM mixture possess components that may assist in the absorption, retard the metabolism, or maintain the stability of the active component(s). In these cases, the active components should be identified by a molecular-based HTS assay.

8. THE FOREFRONT OF MOLECULAR NUTRITION

Utilizing the discovery resource of an intelligently compiled Food Library and the selection of disease-relevant and validated molecular targets, while applying state of the art technologies such as HTS, and the chromatographic and spectroscopic tools of natural products chemistry, is a novel approach in identifying bioactives from food, medicinal plants, and edible materials. The overall research approach described is not unique and applies to many of the discovery and development research strategies currently utilized by the pharmaceutical industry to identify molecules from synthetic compound libraries. The novelty of the described approach lies within the systematic collection and generation of a unique, diverse, and rich Food Library as a discovery resource.

In conclusion, we believe that we are at the forefront of creating a new scientific discipline termed "molecular nutrition". Molecular nutrition will seek to identify previously unrecognized biologically active components in food, medicinal plants, and other edible materials. These bioactives will be capable of maintaining or improving metabolic fitness and thereby preventing or postponing the onset of disease stages in people with borderline disease conditions. This effort will employ rigorous scientific studies of the identified bioactives in the laboratory; *in vivo* animal studies and in humans through appropriately designed randomized, double-blind, placebo-controlled clinical trials for both safety and efficacy. Efforts will be made to identify and standardize these bioactives and understand their biological activity at a molecular target and/or mechanism-based level. This unique research approach will allow researchers to identify and study bioactives from natural sources such as food, medicinal plants, and edible materials, and to make unique discoveries of structure-function relationships, to explain the observed health benefits at a molecular level and to make health benefit claims for these products under DSHEA. As a result, this research will provide consumers and health care professionals with alternative choices to prescription drugs by offering safe, proven, efficacious OTC products containing natural, food-derived compounds for the treatment and prevention of chronic health maladies.

ACKNOWLEDGEMENTS

The authors would like to acknowledge Drs. Ganesh Kishore, Larry Kier, Maureen Mackey, Jim Miller (Missouri Botanical Garden) and Ms. Diane Herdon for many helpful and stimulating discussions.

REFERENCES

Burn, P. & Kishore, G. (2000). Food as a Source of Health Enhancing Compounds. Ag. Bio. Forum, 3, 255-261.

Della Penna, D. (1999). Nutritional Genomics: Manipulating plant micronutrients to improve human health. Science, 285, 375-379.

Kritchevsky, D. (1995). Nutrition and Health, ed. Bronner, F. (CRC, Boca Raton, FL), pp. 89-112.

Mazur, B., Krebbers, E. & Tingey, S. (1999). Gene discovery and product development for grain quality traits. Science, 285, 372-375.

Moffat, A.S. (1999). Crop engineering goes south. Science, 285, 370-371.

Tyler, V.E. (1999). Phytomedicines: Back to the future. J. Nat. Prod., 62, 1589-1592.

Current address:

Dr. Steve Bobzin,
Galileo Labs
5301 Patrick Henry Dr. Santa Clara, CA 95054,
e-mail: scbobzin@yahoo.com

Dr. Paul Burn
 Eli Lilly, Lilly Corporate Center
 Indianapolis, IN 46285
 e-mail: Burn@Lilly.com

Chapter 6

REGIONS IN CHINA RICH IN RESOURCE FOR MEDICINAL PLANTS

A. ANHUI PROVINCE

ZUOJUN JIANG
Vice Governor of Anhui Province
Peoples' Republic of China

1. HISTORICAL BACKGROUND AND TRADITIONAL CHINESE MEDICINE IN ANHUI

There is a long history and tradition in Anhui Province in the research and application of Traditional Chinese Medicine (TCM). More than 1400 experts have worked on TCM over the past several thousand years. The Province is in the center of those provinces rich in herbal resources. To the east, there is Zhejiang herbal medicine; to the west there are Sichuan, Hubei, and Hunan herbal medicines. Anhi Province, itself, hosts a variety of natural medicinal herbs (2700 types) and is first in China in the production of herbaceous peony, *Poris cocos*, chrysanthemum, and the root bark of the tree peony. These herbal medicines are popular in both domestic and international markets. The Bozhou TCM Special Market in Anhui Province is the largest distribution enters in China for trading herbal medicines. The annual trading value is about 6 billion RMB yuan. Through biological technology we aim at modernization, industrialization and globalization.

Yuan Lin (ed.), Drug Discovery and Traditional Chinese Medicine: Science, Regulatory and Globalization, 55-60.
©2001 *Kluwer Academic Publishers. Printed in the Netherlands.*

2. MODERNIZATION OF TCM IN ANHUI

To meet the GAP (Good Agricultural Practice) standard, we intend to set up TCM planting bases such as the one in the city of Wuhu for modernizing the production of TCM. Anhui Province is developing a garden of 200 hectares, in the Wuhu Economic and Technology Development Zone (approved by the State Council in 1993) to be called The Science and Technology of Wuhu Biological Medicine Industry. Wuhu Guangxia Biological Technology Co. Ltd. has imported supercritical fluid extracting equipment from Germany that will be used this year to extract from animals and plants biological ingredients of high purity. This equipment can process 4500 cubic liters of fluid. The use of this technology will enable us to overcome long existing problems: ineffective absorption of medicine (normally over 30% of important ingredients are lost), difficulty in reducing the size of preparations, retention of harmful materials. This method gives us an extracted substance of high purity and clarity, and strong biological activity which conforms to the universally acknowledge features for green and natural medicine. The Science and Technology Garden of Wuhu Biological Medicine Industry is divided into five functional parts: 1. Biological extracting technology center, 2. Natural medicinal research and development center, 3. Biological and pharmaceutical industry, 4. Chinese Medicine GAP culture demonstration center, 5. Biological and Medicine Park. This science garden relies on the abundant resources for herbal medicine in Anhui Province and in East China and on research provided by biological and pharmaceutical enterprises. We welcome talent, technology and investment from interested parties both at home and abroad.

We warmly welcome friends from abroad to visit Anhui Province and the Science and Technology Garden of Wuhu, and we offer the opportunity to others to invest and cooperate with us. Let us make a joint effort to contribute to human health and the modernization, industrialization, and globalization of TCM.

B. HUBEI PROVINCE

LINCHENG MA
Vice Mayor, Jinshou City, Hubei

1. INTRODUCTION TO THE HUBEI PROVINCE

Hubei province, located in the center of the People's Republic of China, has a population of 59 million and an area of 18 million square meters. Hubei is rich in natural resources. The eastern part of the province is the Daibei mountain range, the mid-section is the rich Jianghan plain, and the western section is known for its natural beauty. Hubei is also famous for its tourist attractions, such as Wudang and Ba mountain ranges and the Three Gorges Dam. The largest river, the Yangtze River, divides the province into north and south.

Hubei province is also known for its well-developed transportation network. The nation's largest railroad system, the Jianguang Rail system, runs through the middle of the Hubei province. China's two largest highways, Beijing to Zhuhai and Shanghai to Chengdu, also intersect in the province. In addition there are four major airports in Hubei that provide convenient air travel. Throughout the long history of China, Hebei has been a key military post and is also a center for commerce and trade.

2. THE MEDICINAL PLANT RESOURCES OF HUBEI PROVINCE

The climate of Hubei is mostly subtropical, with the mountainous area having a temperate climate. There is plenty of sunshine and rainfall and the region has a prolonged frost-free period. These climate conditions are particular suitable for the growth and propagation of plant and animal species. Thus, it is not surprising that Hebei has remarkable biological diversity.

Many of the plants and animals found in the Hubei province have medicinal properties. Hubei is one of the provinces in China that has abundant medicinal plant resources. Of the 12,807 known species of medicinal plants and animals found in China, 3,939 species are found in Hubei, including 3,354 plants, 524 animals, and 61 minerals. Hubei ranks fourth in the nation for its different variety, seventh in cultured production for medicinal use, eleventh for wild harvest, and sixth in usage. Recently, 60 species of Chinese medicine were selected for in depth research, including nine species that are indigenous to Hubei. Thus, Hubei is indeed one of the most important TCM resources in China.

The main medical resources of Hubei province are distributed in the Shenglongjia Primeval Forest and its surrounding areas, the Enshi Tujia and Miao nationality autonomous prefecture, the Three Gorges reservoir area, the Dabie mountain area, and the Wudang mountain area. Most of the natural medicine is found in the Wudang, Tongbo, Wu, and Dabie mountain areas. Hubei ranks number ninth in root and rhizome medicine, seventh in seeds and fruit medicine, and ninth in grass medicine. In Hubei, three to five percent of natural production is flower medicine, 10% is leaf medicine, and over 10% is animal medicine.

The Shenglongjia Primeval Forest area has incredible plant and animal diversity. The diversity in the Shenglongjia Primeval Forest is one of the greatest in the world. A recent count in the Shenglongjia area showed that there are 195 families, 816 categories, and 1886 kinds of medicinal vegetables, 111 families, 172 categories and 211 kinds of medicinal animals, 14 kinds of medicine minerals, and 17 other kinds of medicine resources. Of these species, and it is believed that there are more unknown, there are 148 kinds of rare and valuable animals and vegetables.

Enshi Tujia and Miao nationality autonomous prefecture has rich Chinese Medicine resources. This prefecture is regarded as the *Natural Medicine Storehouse in Middle China* from ancient times. In the Enshi area, there are 2,088 kinds of medicinal plants, which is nearly 20% of the number of medicinal plants in China. There are more than 100 kinds of medicinal plants, 120 kinds of medicinal animals, and more than 30 kinds of minerals. There are 282 kinds of Chinese medicine products, including more than 40 kinds of products that are exported to foreign countries.

The Wudang mountain area, which is located in the northwest of Hubei province, is regarded as "the treasure house of medicine plants" from the Ming Dinestry (350 years ago). *BENG CAO GANG MU*, a famous Chinese Medicine book, states that 70% of its animal and vegetable specimens were picked up in the Wudang mountain area. According to the investigation, there are 1,900 kinds of medicine resources in Wudang Mountain. The total reserves are about 800,000 tons, which is about 9.5% of national amount.

3. CONSTRUCTION OF HUBEI CHINESE MEDICINE BASE

Recently, with increasing attention to Chinese Medicine, the development of Hubei's Chinese Medicine resources has completely changed. The variety, structure and distribution of medicine production have become more complete. Some high quality medicine production bases have been constructed. Hubei province has many advantages in the field of Chinese Medicine production, including its diverse topography, its means of distribution, its support from the government, and the existence of numerous famous medicine industries.

3.1 HUBEI'S DIVERSE TOPOGRAPHY AND DISTRIBUTION ADVANTAGES

Hubei's unique geographic position causes the province to have diverse topography. Hubei is a transition area from a temperate zone to a subtropical zone. It has mountains, plain, hills, and a lake, and this variety provides a habitat for many kinds of plants and animals. This area receives plenty of sunshine and rainfall, providing a very good natural environment for the growth of species used to produce Chinese medicine. Additionally, Hubei's developed transportation system makes the province a main production, collection and distribution area for Chinese Medicine.

3.2 HUBEI'S RESOURCE DEVELOPMENT AND UTILIZATION ADVANTAGES

Hubei also has a number of resource development and utilization advantages. In order to establish a competitive advantage, Hubei Provincial Party Committee and Hubei Provincial People's Government has listed the provincial Chinese Medicine resource development as part of the emphasis development area of Hubei province. With support from departments, such as the provincial Economic Commission, and from several research institutes, such as the Hubei Chinese Medicine College, some Chinese medicine production bases have been built in Hubei province.

3.3 HUBEI'S MEDICINE INDUSTRY ADVANTAGES

There are already numerous famous medicine industries in our province, such as the WUHANJIANMING Medicine Company, MANYINGLONG Medicine Company, and several others. Besides these, Tianfa Company and Huoli 28 Company will jointly develop the Chinese Medicine resources of our province. Now a new Huoli 28 High-tech Biology Area, which covers 600 mu, has been built in Jingzhou, Hubei. There is no doubt that this new area will be the starting point for further development of Hubei Chinese Medicine industry.

4. THE CHINESE MEDICINE INDUSTRY OF HUBEI AND THE THOUGHTS OF TIANFA

It is well known that, in the 21st century, Biological Engineering, Synthetic Medicine, and the Natural Plant Medicine will dominate the world medicine industry. Although there are some differences between plant medicine and Chinese Medicine, Chinese Herbal Medicine can be regarded as either one. The extracted plant materials are herbals, grain plants, and other plants.

In addition to the production of Chinese Herbal Medicines, the Hubei province is also the main production area of grain, cotton, oil and vegetable of China. Some examples of the annual output include 20 million tons of rice, 4.5 million tons of wheat, about 300,000 tons of soybeans, two million tons of oil crops, and 100,000 tons of garlic.

Hubei Tianfa Co. Ltd. is a synthetic company that has business in agriculture, industry, and trade. The total assets are over 10 billion RMB and output value is about 6 billion RMB. Two of the companies that are part of Tianfa Co. Ltd., Tianfa and Huoli28, are currently on the market. Since the development of this company, Tianfa has had a lot of operation and management experience. The new area of development will be the field high-tech biological medicine.

The basic idea is to build the largest medicine storehouse of China in Hubei. Within three to five years, the company aims to make the growing area of Chinese medicine over 1 million mu, 100,000 of which must achieve Good Agricultural Practices (GAP) standards. Another goal is to build a national Chinese Medicine development and research institute to do traditional and new resource development research. Tianfa Co. Ltd. also aims to reform some related medicine plants, to build Good Manufacturing Practice (GMP) standard production systems in the plant, and to acquire two or three Chinese Medicine products that have received FDA approval. These steps will lay a foundation for Chinese Medicine to enter new international markets.

We believe that the planning and implementation of the Chinese Medicine storehouse will be a new starting point for the development of the Hubei Chinese medicine industry. Hubei warmly welcomes friends from other countries to visit and inspect Hubei facilities and to invest and build a joint venture. Hubei would like to work with others for the people's health, happiness, and the beautiful life, as well as future development of the Chinese Medicine industry.

Chapter 7

THE CAMPTOTHECIN EXPERIENCE: FROM CHINESE MEDICINAL PLANTS TO POTENT ANTI-CANCER DRUGS

STRINGNER S. YANG, GORDON M. CRAGG AND DAVID J. NEWMAN
Natural Products Branch, National Cancer Institute
National Institutes of Health
Bethesda, MD 20817

Abstract: Camptothecin is a potent natural product based anticancer agent isolated from an organic extract of the bark of a Chinese tree, *Camptotheca acuminata,* Decaisne (Nyssaceae). The chemical identification of camptothecin was first reported in 1966 by Wall *et al.* (1) as a plant alkaloid which showed strong potency against a number of cancer cells derived from leukemia, small cell lung cancer, and colon and rectal cancer. This communication describes the drug discovery and development process of camptothecin from *Camptotheca acuminata* at the NCI in collaboration with Monroe E. Wall's group of the Research Triangle Institute and other laboratories facilitated by NCI support. This led to the eventual development of several potent camptothecin analogs, 9-aminocamptothecin, topotecan, irinotecan, and rubitecan, which have been approved for cancer therapy against diverse cancer types including different leukemias and lymphoma, small cell lung carcinoma, colon and rectal cancer, cancer of the central nervous system, renal cell carcinoma, ovarian and breast cancer in humans. Topotecan and irinotecan are also currently used in a wide spectrum of clinical trials in patients with metastatic cancers. The resolution of camptothecin action mechanism targeting topoisomerase I provided vital information in cancer drug action, and stimulated novel and rational cancer drug design and development.

1. INTRODUCTION

Yuan Lin (ed.), Drug Discovery and Traditional Chinese Medicine: Science, Regulatory and Globalization, 61-74.
©2001 *Kluwer Academic Publishers. Printed in the Netherlands*

1. INTRODUCTION

The history of mankind is well documented with medicinal applications of plants and other natural products. As early as the 2800 B.C., in the *Sheng Nung* era of ancient China, the very early plant derived remedies for curing diseases, were documented in Sheng Nung Ben Cao Chien, and "The Herbal Classic of the Divine Plowman" written by an anonymous scribe in about 101 B.C. based on medicinal records obtained from the royal courts. Throughout the Chinese dynasties, information had accumulated in royal court collections, books and in pharmaceutical records of formulas and/or prescriptions that describing diverse plants employed in herbal medicines, the science of alchemy involving heavy metals and other elements such as zinc, mercury and sulphur, insects and animals, in parts or total, as applied to their therapeutic use in the ancient arts of curing diseases. The best-known example is Mahuang, derived from the dried plant stem of *Ephedra sinica*, constituted principally of the alkaloids, *l*-ephedrine (80-90%) and *d*-pseudoephedrine, methylephedrine and *l*-norephedrine. It is a strong antiasthmatic herb and is widely prescribed for treatment of asthmatic condition. This and many such formulas are currently used in Traditional Chinese Medicine (TCM). A comprehensive English language description of natural products used in TCM can be found in the two volumes of "Pharmacology and Applications of Chinese Materia Medica" edited by Drs Hson-Mou Chang and Paul But (2).

Within the past decade, ethnobotanical products and native herbal shrubs in the form of complementary therapeutic formulas, which have been widely accepted in the East and the Far East, as well as in other traditional medicine systems, have become sought after as an alternative source of medications in the West. Many of the TCM formulas and herbal medicines, such as St. John's Worts and Echinaceae, with presumed medicinal and remedial merits, are widely accepted within the limits of alternative medicine or nutritional supplements. For example, there has been an accumulation of data that pointed to the increased risk of gastric, esophageal, nasopharyngeal and bladder cancers as related to cumulative exposure to N-nitroso compounds (3). The popularity of green tea as well as vitamins C and E, was based on the implication that they might induce urinary excretion of two breakdown products of the N-nitroso compounds, N-nitrosodimethylamine and N-nitrosopiperidine, both of which are known carcinogens in test animals. It was thus presumed that ascorbic acid and moderate intake of green tea can reduce the formation of N-nitroso compounds in the body. Green tea and ascorbic acid are therefore considered as having potential in cancer prevention or lowering the risk of cancer (4). Consumers are increasingly interested in these types of alternative medicine because they are obtainable without prescription, easy to take, and are claimed to have less undesirable side effects as compared to conventional medication. Indeed, TCM does have formulas that are reported to exhibit activity against asthma, metabolic diseases, cardiovascular disease including hypertension and blood pressure irregularities, pain, depression, infectious diseases including AIDS, acting as immune stimulants, as well as cancer. Alternative medication and TCM, thus offer desirability and hope especially against diseases that are recalcitrant to the current state-of-the-art therapy with conventional medication; they presumably also contribute to disease prevention, improvement in health and quality of life among the aging population. According to one market survey reported in 1997, the annual sales of herbal medications along with such nutritional supplements plus vitamins worldwide, exceeded 3 billion U.S. dollars in Taiwan alone, and approached 16 billion dollars worldwide. A more recent overview reported in the Nutritional Business Journal documented that, in 1999, nutritional business revenues from sales of natural and organic goods, supplements, functional foods and personal care products skyrocketed to 44.5 billion dollars in the U.S. (5). Sales of herbal products including soy products, Ginseng, green tea, Echinacea, Ephedra, St. John's Wort, Ginkgo biloba, Garlic, Saw palmetto, as well as glucosamine, to name a few, were seen rising in 1997 and 1998 to a total of over 2.3 to 2.6 billion U.S. dollars respectively (6).

As we embrace herbal products and TCM as alternative medicine or "food supplement" in the western hemisphere, we need to confront the real issues involved in the making of a therapeutic agent. This includes definitions in chemistry, standardization, preclinical testings, *in vivo* testings, pharmacology, toxicology, bioavailability, formulation and quality control. The discovery and development of camptothecins illustrate that these seemingly complex technical challenges can be successfully applied to the production of genuinely effective agent for cancer therapy.

2. THE DISCOVERY OF CAMPTOTHECIN FROM *CAMPTOTHECA ACUMINATA*, DECAISNE (NYSSACEAE), A MEDICINAL TREE NATIVE TO CHINA, AS AN ANTICANCER AGENT

Camptothecin is a natural product based anticancer agent first identified in an organic extract prepared from the bark of a Chinese tree, *Camptotheca acuminata,* Decaisne (Nyssaceae). *C. acuminata* has been known as a medicinal tree in China; it is native to the Shih-Fang Hsien. As early as May 19, 1908, it was reported by E.H. Wilson of the U.S. Department of Agriculture. The tree can also be found in other locations of China including Szechwan's Red Basin, Chengtu, Chungking, Kiukiang, Yunnan and Kwangsi

(7). *Camptotheca acuminata* is unique to China since it has not been identified anywhere else in the world. The Szechwan's Red Basin, surrounded by mountain ranges, is considered an ideal botanical conservation area with many unique species that survived the abrasive weather actions during the ice age because of its geographical advantage; this was superbly described by Dr. Raven's lecture in this symposium (8). In the early 1950's the United States Department of Agriculture (USDA) was interested in collecting plants with certain phytochemical resources, such as steroidal sapogenins. *Camptotheca acuminata* was first introduced into the U.S. by the DOA in 1911 from Mt. Omei, Szechwan. By January 12, 1927, 81 shipments were made to the Washington, D.C. by W.T. Swingle and distributed among the botanical gardens on both U.S. east and west coasts for cultivation. Culturing of the Nyssaceae species of *Camptotheca acuminata* Decaisne was carefully planned by botanists of the DOA, matching climatic conditions between the native environment and the U.S. selected locations, as well documented in a DOA technical bulletin (No. 1415) (7).

In the 1950's DOA was looking for steroidal sapogenins, but instead they found flavonoids, tannins and sterols in the extracts of *C. acuminata* roots, bark, woody stems, leaves, flowers and seeds. In 1961, aqueous ethanol extracts of bark and wood, and fruits of *C. acuminata* from the Chico Station, California, prepared and submitted by Monroe Wall's group of Research Triangle, N. Carolina, were found active against diverse tumor cells including L1210 lymphoid leukemia cells, adenocarcinoma 755 (Ca 755), and sarcoma 180 (Sa 180), as well as other tumors, such as Lewis lung carcinoma and human sarcoma in the CCNSC screening program of the NCI (1, 7, 9). In general, it was reported that the antitumor effect of *C. acuminata* was greater when the extract was prepared from root, than from bark and stem, and least from fruits and leaves. Tree age made no difference in the relative abundance in the yield of (CPT[1]) as long as the trunk diameter exceeded 1/4 inch. Camptothecin was found to concentrate in the wood. Maximum yield of camptothecin was obtained with trees producing maximum dry matter (7). By 1963, an enhanced effort supported by the National Cancer Institute, enabled Wall's group to isolate significant quantity of camptothecin from *Camptotheca acuminata* Decaisne for chemical analysis of the molecular entity that exhibited the anti-tumor activity. The method of isolation involved

---

[1]CPT was an acronym given to Camptothecin in 1970's. Later CAM was also used as an acronym for Camptothecin.

sequential ethanol extraction, chloroform phasic countercurrent distribution fractionation, then followed by purification using silica gel column chromatography and eventual crystallization. The procedure constituted a gentle but effective way of producing pure camptothecin in significant quantity (1, 9). By 1966, the structure of camptothecin as a novel alkaloidal antileukemic and anti-tumor inhibitor was established and documented (Fig. 1)(1, 7, 9).

3. CHEMICAL STRUCTURES OF CAMPTOTHECIN AND ITS ANALOGS

The chemical identification of camptothecin (CPT) (was first reported in 1966 as a plant alkaloid with a molecular weight of 348.11 dalton, $C_{20}H_{16}N_2O_4$ (1). Camptothecin (NSC 94600 and 100880)[1] showed strong potency against a number of cancer cells when tested against the NCI 60 cancer cell line panels. Figure 2 shows the mean graphs of the inhibitory activities of camptothecin, NSC 100880, obtained in a typical screening assay against diverse cancer cell lines carried out in late 1990's. Mean graphs are the optimized graphic presentation of a composite of data obtained from the NCI *in vitro* cell line screen. It provides an unique capability of comparing the relative drug efficacy against the different cancer cells within the same plot (10). To put it simply, an extension of the bar towards the right of the mean (at 0) indicates the molarity obtained for a drug to which that particular tumor cell line is sensitive, when compared to the averaged control values (MG-MLD). In short, it also indicates the $\log_{10}$ molarity at which the drug kills the tumor cell at the indicated level, which is the same numerical value given on the left of each bar. It is evident from figure 2, that camptothecin is effective in the *in vitro* assays against different leukemia, non-small cell lung cancer, central nervous system cancer and renal cell carcinoma cell line, albeit the relative effectiveness differs dependent on the cancer cell type.

---

[1]NSC number is the numerical code given to a compound when it is submitted to the NCI for anticancer or anti-HIV screening. Camptothecin was first given the NSC 94600 in early 1966, and later given NSC 100880 for the 20(S)-camptothecin preparation.

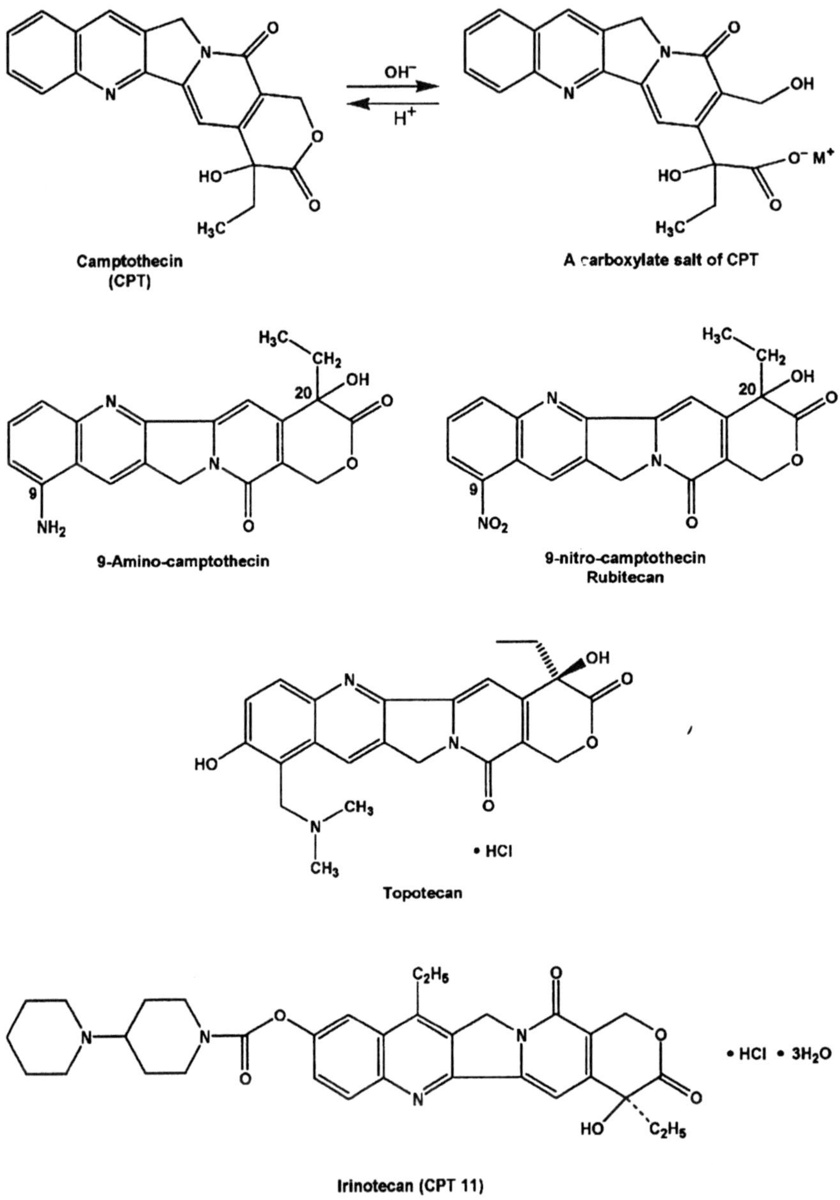

Figure 1.   Chemical structures of camptothecin and its analogs.

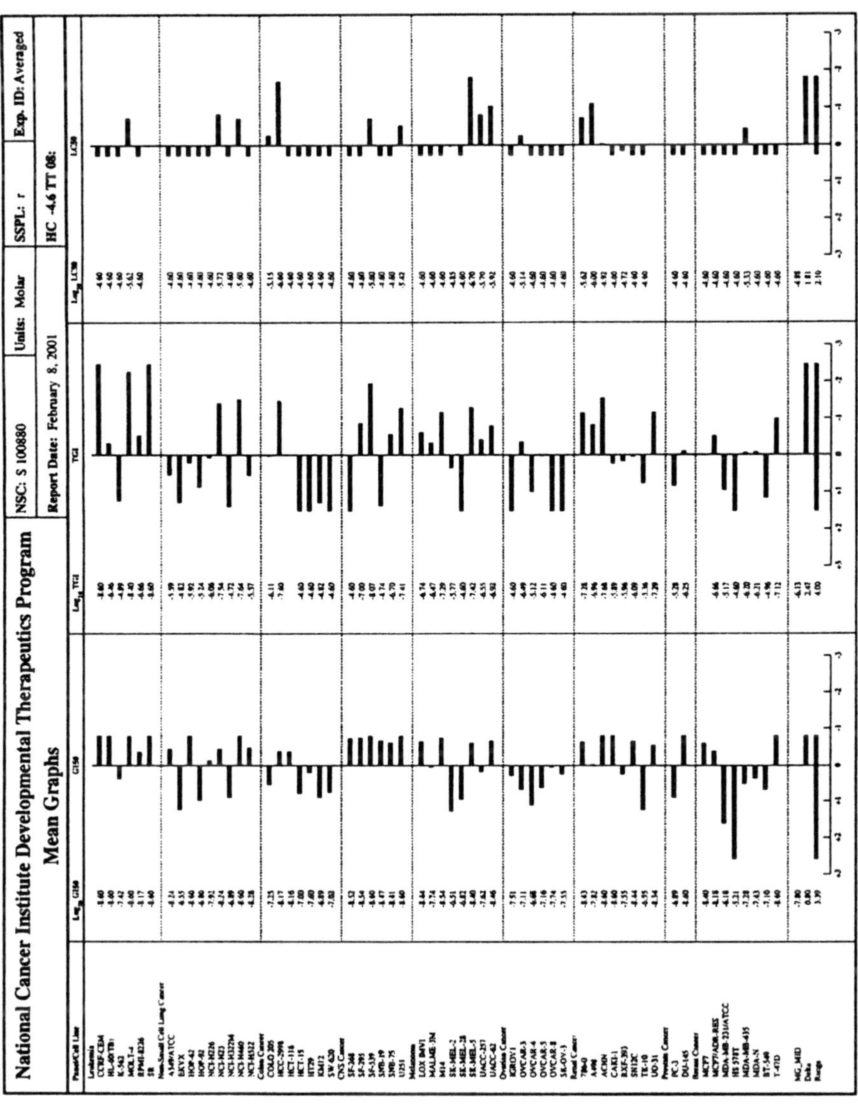

Fig 2. Mean graphs depicting the inhibitory activities of camptothecin (NSC 100880) against diverse cancer cells in the 60-cell line anticancer drug screen of the NCI.

68

Camptothecin was found to be relatively insoluble in water and in many organic solvents with the exception of dimethylsulfoxide. Led by Wall's group in the Research Triangle Institute and other laboratories supported by NCI, efforts in late 1960's to early 1970's focused on developing CPT analogs (Figure 1) that will overcome the solubility problem. A great many analogs with substitutions in the 9, 10, or 10,11- positions were made (9, 11). The current effective CPT analogs are the 9-amino- and 10,11-methylenedioxy-CPT derivatives, CPT-11, and topotecan, that showed strong efficacy against human colon cancer xenografts in nude mice (11). 9-nitrocamptothecin was developed in late 1990's and can be considered as a prodrug of 9-aminocamptothecin since the 9-nitro group is converted to 9-amino group in the body. 9-nitrocamptothecin is still in clinical trial whereas the other camptothecin analogs are now available for public use as well as in a wide spectrum of clinical trials as discussed below.

4. EARLY STUDIES ON ANTITUMOR EFFECT WITH CAMPTOTHECIN AND CLINICAL TRIALS

In late 1960's and early 1970's, camptothecin was found to be readily converted into a sodium salt with sodium hydroxide, or an acetate and a chloride and iodoacetate by treatment with the appropriate sodium chloride or sodium iodide in acetone Figure 1 (9). The sodium salt of camptothecin increases its solubility and thus easing delivery of the drug. The NCI, encouraged by the broad spectrum antitumor activity of camptothecin observed in animal testing results, proceeded to two Phase I clinical trials with the sodium salt of camptothecin in 18 patients (12, 13). Only partial responses in 5 out of 18, and 2 out of 10 patients were observed. Responses were seen in patients with gastrointestinal tumors and only for a short duration. This observation corroborated with the results of a separate phase II study with 61 patients, suffering from adenocarcinomas of the gastrointestinal tract. In the latter phase II study, only two patients out of 61 patients showed objective partial responses. It was then realized that CPT sodium salt exhibited only one tenth of the anti-tumor activity observed earlier and that the hydrolysis of the lactone in ring E in mild alkali solution, accounted for the lowering of the antitumor activity observed with the CPT sodium salt. Confronting the unexpected diminished response, patients were administered continuously higher doses of CPT sodium salt so as to maximize the effectiveness of CPT (13, 14). This led to an accumulation of CPT in the bladder due to the subsequent relactonization of CPT sodium salt in the urine acid environment, resulting in severe hemorrhagic cystitis besides other toxicities involving myelosuppresion, vomiting, diarrhea, bladder ulceration, which were quite intolerable to the patients, inevitably forcing the discontinuation of all phase II trials.

Camptothecin was also tested in the People's Republic of China with 1000 patients. Results were more encouraging as seen in some positive results reported in gastric cancer, intestinal cancer, head and neck tumors and bladder carcinoma (15). Toxicities observed up to this time are limited to dose-limiting hematological depression, vomiting and diarrhea with no serious threat to the patients.

Confronting such complications in solubility, formulation and drug delivery problems, interest in camptothecin waned in mid 1970's to early 1980's.

5. MECHANISM OF ACTION OF CAMPTOTHECIN ANALOGS AND THE PRODUCTION OF EFFECTIVE POTENT ANTICANCER DRUGS

In the early 1970's camptothecin was already known as a S-phase inhibitor, exerting very early effects in the cell cycle (16-18). It interfered with several key steps on the cellular level. A more recent report described the cytological changes in the nuclear morphology of HL-60 and L1210 leukemic cells after 3 to 6 hour treatment with CPT at therapeutic levels (16). Nuclear chromatin fragmentation or "pulverization" and nuclear enlargement with irregularity in shapes of HL-60 cells took place when compared with control cells. And, in the L1210 cells, an increment of micronuclei and nuclear enlargement also took place after CPT treatment.

Interest in camptothecin rekindled upon the discovery of its anti-tumor action mechanism. In 1985, Liu *et al.* identified the molecular target of CPT as DNA topoisomerase I, a nuclear enzyme pivotal to DNA replication and transcription (16, 17). Biochemically, camptothecin was reported to inhibit DNA synthesis in an irreverisble or partially reversible manner, but it inhibited RNA synthesis in a highly reversible manner. The significance of CPT anti-tumor activity became self-evident when it was found that CPT inhibition involved the binding of DNA topoisomerase I at a critical point during DNA synthesis, resulting in an accumulation of protein-linked DNA breaks (17). Figure 3, based on evidence reported by Leroy F. Liu and others (16-20), depicts the sequential steps of CPT inhibition of topoisomerase I catalysis during DNA replication. As the double-stranded DNA unwinds to

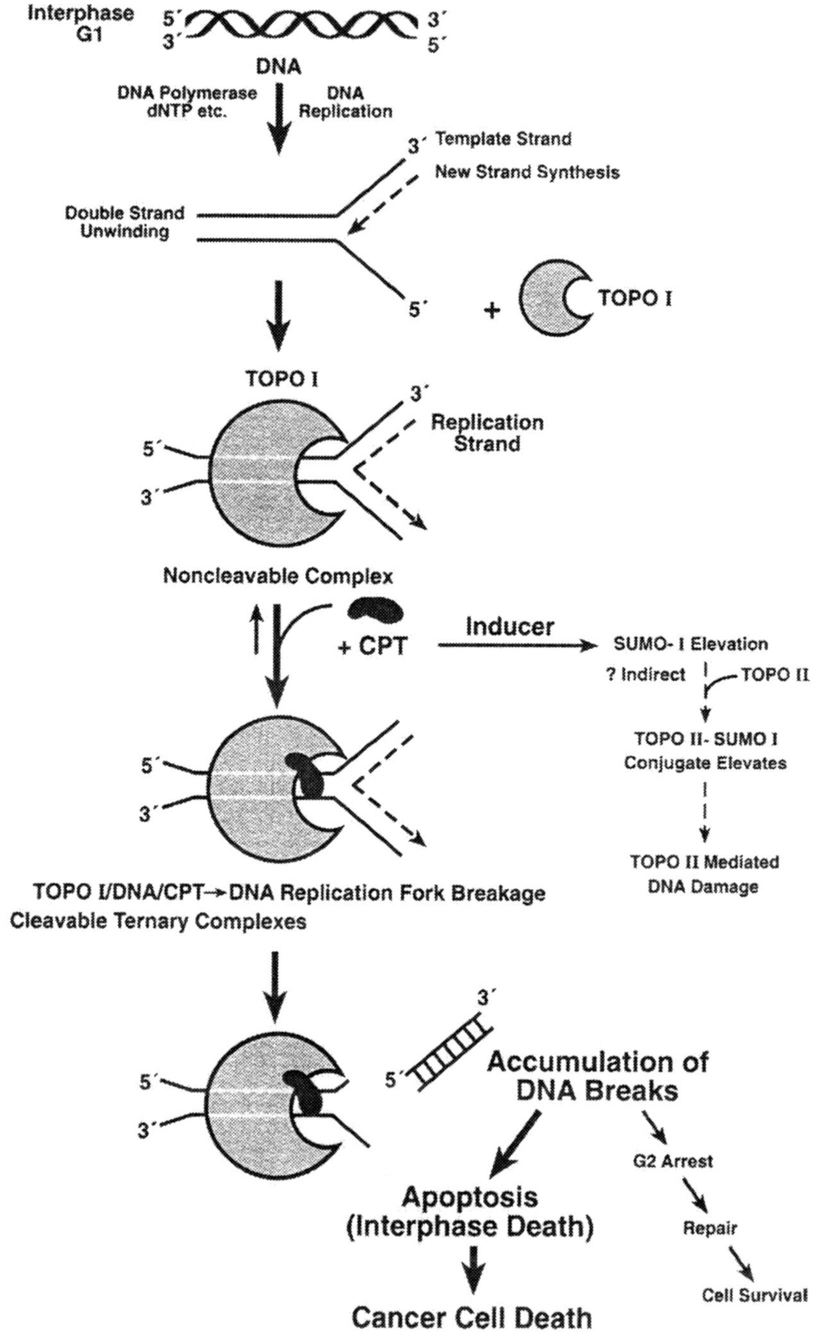

Figure 3. A scheme of possible action mechanism of camptothecin targeting topoisomerase I during DNA replication in cancer cell.

permit *de novo* replication of the single strand, the process involves a breaking and then a rejoining step of the newly synthesized DNA strand catalyzed by DNA topoisomerase I. CPT inhibition targets specifically at the rejoining step. It stabilizes the cleavage complex, thus causing an accumulation of the covalent reaction intermediate, i.e. topoisomerase I-CPT-DNA ternary complex and generating double stranded DNA breaks. This leads to an accumulation of double stranded DNA breaks and an arrest of progression into the G2 phase of the cell cycle, and in turn, triggers the possible events that lead to a cascade of cellular events, i.e. apoptosis, besides other cellular responses, and eventual cell death.

Recent detailed studies on DNA sequence analysis of topoisomerase I cleavage sites induced by camptothecin (20-21) led to the recognition of other drugs with similar action (22). This further documents the significance of the camptothecins in recent history of anticancer drug discovery and development.

6. CURRENT CLINICAL TRIALS WITH CAMPTOTHECIN ANALOGS

Because of CPT's ability to interfere with such fundamental cellular function, its anti-tumor characteristic is therefore broad spectrum. Also, with the revelation of the importance of the lactone ring in preserving its inhibitory action, heightened research efforts in the 1980's and early 90's concentrated on formulation, pharmacology and drug delivery, as well as molecular modification at positions 7, 9, and 10 of the parent 20(S)-camptothecin compound, so as to optimize CPT delivery within maximum tolerable dosage in cancer therapy. The four most potent and widely used camptothecin derivatives all possess intact lactone ring structure (Figure 1, Table 1). Currently there are over 130 clinical trials involving the most potent CPT analogs, 9-aminocamptothecin, CPT-11, topotecan, and 9-nitrocamptothecin in therapy of diverse cancers at either early stage or at late metastatic stages (Table 1). There were a great many important studies that led to the development of these four CPT analogs and their approval by the FDA for patients' use. Many of such studies can be found in the three symposia publications on camptothecins (23-25). CPT-11 is currently produced and marketed by Pharmacia Upjohn under the trade name of camptosar or irinotecan. 9-nitrocamptothecin, developed by Yakult Honshu and Daiichi Seiyaku, was licensed to SuperGen in 1997, and is currently given the trade name of rubitecan. Rubitecan can be considered a prodrug of 9-aminocamptothecin since the 9-nitro group is converted to 9-amino group when absorbed by the body. It is still at the clinical trial application stage. The information listed in Table 1 has been streamlined and simplified to list primarily the different cancer types in these clinical trials. Information presented in Table 1 by no means try to detail the clinical trials currently in progress. Details of the individual cancer trials can be readily obtained by accessing the cancer trial web site of the CancerNet of NCI at the following Internet address: *"http://cancernet.nci.nih.gov/"*. One important feature not highlighted in this table is the use of these potent camptothecin analogs in late stage metastatic cancers or in cases with multiple cancers. This type of information can be gleamed from the diverse clinical trials listed at the CancerNet. That camptothecin analogs target the critically pivotal step in the DNA replication process, makes them successful drugs against diverse cancer types.

Table 1. Clinical Trials with Camptothecin Analogs

Cancer Type	Trials	9 amino campto-thecin	Topotecan	Irinotecan	Rubitecan
Colon and Rectal Cancer	21			XXX*	X
Epithelial ovarian cancer or cancer of the Peritoneum	14		XXX	X	X
Retinoblastoma/rhabdomyosarcoma/neuroblastoma/glioma/ brain tumor	14		X	XXX	
Leukemia/myeloid leukemia/multiple myeloma	11		XXX		
Lymphoma/non-Hodgkin's lymphoma/Hodgkin's diseases	7	X	XXX	XX	
Esophagus/stomach cancer/cancer of the pancreas and other gastrointestinal cancer	9	X		XXX	XX
Advanced non-small cell lung cancer/lung cancer	6		X	XXX	
Advanced or recurrent breast cancer	4		X	X	X
Renal pelvis and ureter cancer	1			X	
Others including prostate, Leiomyosarcomas, melanoma	19				
Advanced solid tumors - all types	22		XX	XXX	X
Soft tissue sarcoma	2		X		X
Total	130	2	47	66	15

*The "X" indicates the relative frequency of that analog being used in the clinical trials involving that particular cancer type.

7. CONCLUSION

There is no doubt that the discovery and development of camptothecin provided new insights and perspectives in drug discovery and development from natural products. The camptothecin experience certainly opens new avenues to therapeutic exploitation of natural resources such as those of herbal medicinal extractions.

ACKNOWLEDGEMENT

The author would like to express her appreciation of the very helpful discussion provided by Dr. Leroy F. Liu on the mechanism and update of camptothecin inhibitory actions.

REFERENCES

1. Wall, M.E., Wani, M.C., Cook, C.E., Palmer, K.H., McPhail, A.T. and Sim, G.A.. **1966**. Plant antitumor agents. I. The isolation and structure of camptothecin, a novel alkaloidal leukemia and tumor inhibitor from *Camptotheca acuminata*. J. Am. Chem. Soc. **88**:3888-3890.
2. Chang, H.M. and But, P.P. (Ed.) **1986**. Pharmacology and Applications of Chinese Materia Medica vol. 1-2. World Scientific Publishing Co. Pte Ltd. Philadelphia, PA.
3. Mirvish, S.S. **1995**. Role of N-nitroso compounds (NOC) and N-nitrosation in etiology of gastric, esophageal, nasopharyngeal and bladder cancer and contribution to cancer of known exposures to NOC. Cancer Lett. *93*: 17-48.
4. Vermeer, I.T.M, Moonen, E.J.C, Dallinga, J.W., Kleinjans, J.C.S. and van Maanen, J.M.S. **1999.** Effect of ascorbic acid and green tea on endogenous formation of N-nitroso-dimethylamine and N-nitrosopiperidine in humans. *Mutation Research,.***428**: 353-61.
5. NBJ's Fifth Annual Overview of the Nutrition Industry. 2000. In Annual Industry Overview 2000. *Nutrition Business Journal.* Vol. V, No. 7/8. Pp.3-
6. Bailey, G. M. **2000**. Botanicals: State of the Industry. Pp. 41-
7. Perdue, R.E., Jr., Smith, R.L., Wall, M.E., Hartwell, J.L. and Abbott, B.J. **1970**. *Camptotheca acuminata* Decaisne (Nyssaceae). Source of Camptothecin, an antileukemic alkaloid, U.S. Department of Agriculture, Agricultural Research Service, Technical Bulletin, No. 1415.
8. Raven, P.H. **2001**. Flora, Plant Conservation and China's Future. In Lin, Y. ed. Traditional Chinese Medicine: Science, Regulation and globalization. Kluwer Publishers. (in press)
9. Wall, M.E. and Wani, M.C.. **1995**. Camptothecin and Analogs: From Discovery to Clinic. In *Camptothecins: New Anticancer Agents*, ed. Potmesil, M. and H. Pinedo,CRC Press, Boca Raton. Pp. 21-42.
10. Boyd, M.R. and Paull, K.D. **1995**. Some practical considerations and applications of the National Cancer Institute in vitro anticancer drug discovery screen. *Drug Development Research 34*: 91-109.

11. Wani, M.C., Ronman, P. E., Lindley, J.T. and Wall, M.E. **1980**. Plant antitumor agents. 18. Synthesis and biological activity of camptothecin analogs. *J. Med. Chem.* **23**: 554-560.

12. Gottlieb, J.A. and Luce, J.K.**1972**. Treatment of maglignant melanoma with camptothecin (NSC-100880). *Cancer Chemother. Rep.* Part I, 56: 103-105.

13. Muggia, F.M., Creaven, P.J., Hanson, H.H., Cohen, M.C.and Selawry, O.S. **1972**. Phase I clinical trial of seekly and daily treatment with camptothecin (NSC-100880): Correlation with preclinical studies. *Cancer Chemother.* Rep. 56: 515-521.

14. Stehlin, J.S. **1996**. Introduction and Overview. The Camptothecins: From Discovery to the Patient. Eds. Pantazis, P., Giovanella, B.C. and Rothenberg, M.L.. *Annal of The New York Academy of Sciences.* 803: ix-x. New York, N.Y.

15. Xu, B. **1980**. In **U.S.-China Pharmacology Symposium**. Burns, J. J and Tsuchiatani, P. J. Eds.: *National Academy of Sciences,* Washington, D.C. p.156.

16. Giovanella, B.C., Wall, M.E., Wani, M.C., Nicholas, A.W., Liu, L.F., Silber, R. and Potmesil,M. **1989**. Highly effective DNA topoisomerase-I targeted chemotherapy of human colon cancer in xenografts. *Science* **246**:1046-1048.

17. Hsiang, Y. H., Hertzberg, R., Hecht, S., and Liu, L.F. **1985**. Camptothecin induces protein-linked DNA breaks via mammalian DNA topoisomerase I. *J. Biol. Chem.,* **260**: 14873, 1985.

18. Del Bino, G., Lassota, P. and Darzynkiewicz, Z. **1991**. The S-phase cytotoxicity of camptothecin. *Exp. Cell Research* **193**: 27-35.

19. Ryan, A.J., Squires, S., Strutt, H.L. and Johnson, R.T. **1991**. Camptothecin cytotoxicity in mammalian cells is associated with the induction of persistent double strand breaks in replicating DNA. *Nucleic Acids Research.* **19**: 3295-3300.

20. Jaxel, C., Capranico, G., Kerrigan, D., Kohn, K. W., and Pommier, Y. **1991** *J. Biol. Chem.* **266**: 20418-20423.

21 Pommier, Y., Kohlhaen, G., Kohn, K.W., Leteurtre, F., Wani, M.C., and Wall, M.E. **1995** *Proc. Natl. Acad. Sci. U.S.A.* **92**: 8861-8865

22. Sim, S. P., Pilch, D. S., and Liu, L. F. **2000** Site-specific topoisomerase I-mediated DNA cleavage induced by nogalamycin: a potential role of ligand-induced DNA bending at a distal site. *Biochemistry* **39**: 9928-9934.

23. Potmesil, M. and Pinedo, H., ed. **1995**. **Camptothecins: New Anticancer Agents** CRC Press, Boca Raton.

24. Pantazis, P., Giovanella, B.C. and Rothenberg, M.L. eds. **1996, The Camptothecins: From Discovery to the Patient.** *Annal of The New York Academy of Sciences.* **803**: ix-x. New York, N.Y.

25. Kinghorn, A. D. and Balandrin,M. F. **1992, Human Medicinal Agents from Plants.** ACS Symposium Series 534. American Chemical Society, Washington, D.C.

Chapter 8

APPROACHES FOR EVALUATION OF IMMUNE-MODULATING AND ANTI-TUMOR BIOACTIVITIES IN CHINESE MEDICINAL HERBAL EXTRACTS

PEI-FEN SU, SHENG-YANG WANG, CHIH-CHIEN HSU, SHOW-JANE SUN, YUAN LIN, PEI-LING KANG, CHIN-JIN LI, JOANNA LIANG, LIE- FEN SHYUR*, AND NING-SUN YANG*
Institute of BioAgricultural Sciences, Academia Sinica, Taipei, Taiwan, R. O. C.
(* Corresponding authors)

Abstract: A recent surge of world attention on research in herbal medicinal plants as a potential fast renewable source for new pharmaceuticals or nutraceuticals has attracted many scientists from diverse disciplines to join this scientific venture. These include molecular biologists, bio-organic chemists, pharmacologists, clinicians and traditional Chinese medical doctors. Hence a trend of true multi-disciplinary collaboration in herbal medicine research is being formed.

1. INTRODUCTON

Among the new trends of relevant research approaches, we have chosen two areas of investigation, namely the effect of specific herbs on promoting immune responses and their potential anti-tumor or cancer-prevention activities. For many herbal medicinal plants and their composite formulas, there is in general a lack of modern scientific proof on specific bioactivities, even though they have been often anecdotally referred to as "highly effective" through testimonials of traditional Chinese medical doctors or their ex-patients. Therefore there is an urgent need to scientifically address this difficulty by employing reputable, updated experimental biology systems.

Following a general direction in addressing immune augmentation, we have established, for both mouse and human systems, an *ex vivo* dendritic cell (DC) primary culture system for systematic evaluation of the effect of crude plant extracts and their derived bio-organic fractions from several herbs of our

Yuan Lin (ed.), Drug Discovery and Traditional Chinese
Medicine: Science, Regulatory and Globalization, 75-82.
©2001 *Kluwer Academic Publishers. Printed in the Netherlands*

interest. DC systems have been chosen for they're recently recognized multifaceted involvement in both cellular and humoral immune responses. Dendritic cells can function as an antigen-presenting cell, be responsive to potent cytokines such as GM-CSF and 1L-4, and secret IL-12 which is a key TH1-type cytokine. The dynamics and kinetics of DC maturation also provides an excellent system for fine-tunable experiments in herbal medicine studies, searching for immune modulators and other related effective bio-modifier molecules. This and other approaches for evaluating herbal extracts on immune-modulation activities will be compared. Results of our preliminary studies on the effect of three herbs and a multi-herb formula prescription will be presented.

For assessment of potential anti-tumor activities, cell type-specific cytotoxicity was evaluated with four target human or mouse tumor cell lines, using MTT and /or ^{3}H-thymidine incorporation assays. These activities were further defined and characterized by the apparent involvement of specific signal transduction pathways of apoptosis. Results of this study suggest that a combination of the two above-mentioned systems may provide useful ex vivo tools for fast, preliminary evaluation of immune-modulation and anti-tumor bioactivities in herbal plant extracts.

2. IMMUNE-MODULATION AS A SYSTEM FOR EVALUATING TRADITIONAL CHINESE MEDICINE (TCM) EFFECT

Two very popular indigenous medicinal plants in Taiwan have been selected as the target plants for this study in our laboratories. These two herbal plants have been claimed by local Chinese people for many bioactivities, including the enhancement of immune systems, liver-protection, anti-tumor and others. The purpose of the present study is to use the advanced cellular, biochemical, molecular, and immunological tools to address a number of specific questions on traditional Chinese medicines. First, we heard routinely that Chinese herbal medicines are prescribed on a holistic basis; hence they are not just addressing one specific killing, inhibitory or nourishing effect. "Bu-Qi" is one of the functions, which has been claimed for Chinese medicine. We may interpret that the function of "Bu-Qi" is to keep (enhance) the functions of homeostasis, erythropoiesis, and/or other vital systems such as the immunological defense systems. However, what is the meaning of "immunological defense"? We have heard that "Dong Chong Xia Cao", which is the dried fungus *Cordyceos sinensis* grown on the larvae of the caterpillar *Hepialus armoricanus*, a very popular on commercial natural product in Taiwan and also in China can increase our immune system. But in modern biology terms, we still do not understand the real meanings of increasing our immune system activities by using these herbal extracts. There are thousands of papers that have been published, and are very specifically addressing the questions on specific compounds for modulating specific immune responses, cellular and biochemical responses, etc., but minimal studies have been systematically executed for research into Traditional Chinese Medicines (TCM). Therefore, we need to honestly and respectfully address those observations, and try to merge and use new cellular immunology tools and informations for scientific interpretations of TCM.

3. USE OF DENDRITIC CELL (DC) AS A REPORTER IMMUNE CELL SYSTEM

What are some of the effects of these wonderful "holistic drugs"? We may need to ask these questions with a specific approach: what are the key very important findings during the past 5 years of immunology? In our best observation, we believe it is the dendritic cells (DC) and their involvement in mediating immune responses. DC is is a unique class of antigen-presenting cells (APC) of the immune systems. One of the functions of our immune surveillance system is to effectively recognize the foreign immunogens such as pathogens, and deliver the specific features of these immunogens to the various immune-responsive cells, and via activation of various immune cell activities to eventually eradicate the invading pathogens.

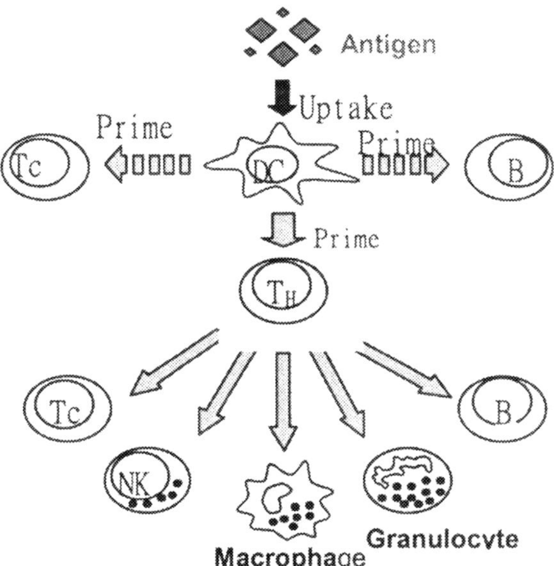

Fig 1. The interaction of DC and the responsive cells and the role of T_H cells in immune systems

DCs are the only one type of APC that can induce such primary immune response and establish immunological memory. DCs are very important for a spectrum of immune activities, not only because for the induction of both primary immune responses and immunological tolerance, but also for the regulation of various T cell-mediated immune responses (Th1/Th2).

The different research fields of immunology have been intensively studied during the past 20 years, researchers have investigated a variety of topics including complement, inflammation, cytokines, T cell-dependent immunity, immunoglobulin, antigen presenting, natural killer (NK) cells, major histocompatibility complex (MHC), hematopoitic stem cells, and others. Recently, every three years or so the technical skills would change and some new cellular immune factors would be identified, so we need to appreciate these new findings and make good use of these emerging new research systems. Dendritic cells are involved in the very upstream of a variety of immune responses, and can serve as the bridge between innate and adaptive immunities. Therefore, we value it is very important to evaluate the possible effects of specific herbal extracts on immune systems by investigating the effect on DC.

The progenitors of DC apparently are originated from bone marrow, they thus move into the circulating system as precursor cells. The circulating DC can home in to non-lymphoid tissues (*e.g.*, as skin), where they can reside as specific types of immature cells (e.g., Langerhan cells) with high phagocytic capacity. After antigen capture, immature DC can process the antigen, and migrate to draining lymphoid organs/tissues where, after maturation, DC can activate naive CD4+ T cells. CD4+ T cells can recognize specific antigens in the context of MHC class II presented on DC and thereby initiate various T cell-mediated immune responses. (For detailed information, see Ref 3)

At least two subsets of DC precursors have been shown to present in peripheral blood, and one of them is the CD14+ monocytes. Many cytokines have been identified and shown to support *in vitro* differentiation and maturation of DC CD14+ monocytes isolated from peripheral blood and cultured with granulocyte-macrophage colony-stimulating factor (GM-CSF)

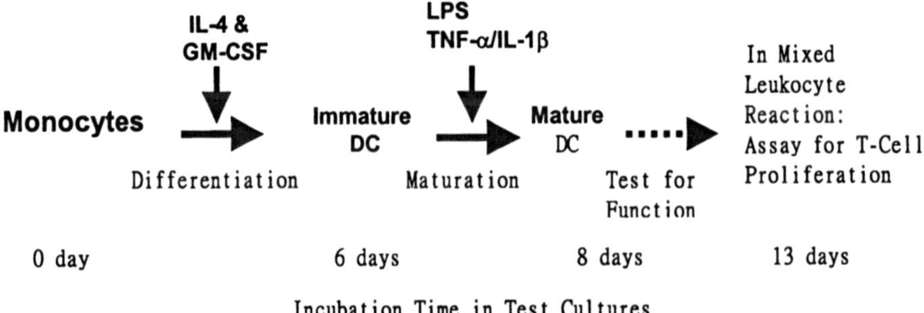

Fig. 2. A strategy to investigate the effect of herbal extract on differentiation, maturation and functions of DC

and interleukin (IL)-4 can effectively undergo differentiation. The resultant cells can behave as efficient APC with morphology and cell surface molecules (CD1a+, CD14-) expressing typical characteristics of immature DC. Tumor necrosis factor (TNF-α), lipopolysaccharides (LPS), and IL-1 β can individually induce further DC maturation and significantly enhance the capacity of DC to stimulate resting allogenic T cells by increasing the expression of co-stimulatory molecules (CD40 and CD 80/86) and specific cytokines. For mature DC, the expression of specific maturating Ag (CD83) is increased and the antigen capture and processing ability of DC are drastically reduced. Since specific cytokines are involved in the differentiation, maturation, and immunological function of DC, we thus asked the question whether specific herbal extracts and their derived phytochemicals can confer measurable and significant effect on these activities. The next question we asked was what kind of herbal extraction strategy we should employ for this task. Most herbal extracts employed in Chinese medicine are cooked and extracted with hot water, and others often are soaked in wine or liquor. We thus have applied these strategies in our experimental system; the herbal plant materials were therefore extracted with cold or hot water and then/ or with ethanol.

To obtain mature DC from monocytes from primary cultures, we have used GM-CSF and IL-4 to induce CD14+ cells for differentiation into a typical immature DC population, which contains a decreased level of CD14 and possess the ability of phagocytosis. After maturation with LPS treatment in culture, the mature DC has exhibited increased levels of CD80, 83, and 86, and was able to stimulate the proliferation of allogenic T cells. In the presence of testing herbal extracts, we have found that specific herbal extract I can effectively influence the differentiation and a specific function of testing human DC samples. Recently, a synthetic immune-modifier has been reported to augment the maturation of DC. In contrast, IL-10, glucocorticoid, and vitamin D_3 were found to block the differentiation and function of human DC. For instance, vitamin D_3 alters the expression profile of the surface markers on immature and mature DC. Moreover, vitamin D_3 enhances the phagocytotic activity of immature DC but reduces the T cell priming ability of mature DC. On one hand, the blockage of DC differentiation and maintenance of the cells at the macrophage-like stage may be able to improve the phagocytotic activity of maturing DC cells, and may thus help eradicate invading pathogens as microorganisms. On the other hand, mature DC has been shown not only to activate CD4+ T cells but also to stimulate Ag-specific CD8+ T cells, thus enhancement of the maturation of DC may effectively improve the function of T cell-mediated immunity. Therefore, it seems very important to study the effect of specific herbal extracts on DC, whether or not they may induce or suppress the differentiation and function of test DC populations. Initial results obtained have in general presented in the conference, detailed results from systematic studies are being prepared for publication elsewhere.

4. ANTI-TUMOR/CANCER PREVENTION ACTIVITIES

We have also performed experiments on a different type of medicinal herb by using hot water for plant extraction. This herb has been a very popular empirically used traditional medicine in treating a number of human diseases and being increasingly used as a popular nutraceutical in Taiwan and other Asian countries. It has also been used as a drink of tea, and is claimed for its liver-protection activity and also for anti-tumor activity. Unfortunately, little or no systematic research has been reported to address these specific pharmacological or nutritional claims of this plant. For evaluation the folklore pharmacological activities of this herb, we first carried out a cytotoxicity study on specific tumor cell lines, including MCF-7 (a human mammary adenocarcinoma cell line), HepG2, B16 melanoma, and other cell lines. We first did the standard MTT and ^{3}H-thymidine incorporation assays. The result showed that a very significant and dosage-dependent cytotoxic effect on MCF-7 cells when they were treated with a specific fraction derived for the crude hot water extract of our testing target herb. A partially purified bio-organic fraction was shown to exhibit a lowest ED_{50} value (Effective Dosage for inhibiting 50% of cell proliferation) of the cytotoxicity on test tumor cells, indicating some of the compounds in this partially purified fraction can confer a much increased cytotoxicity as compared to that of total crude extract.

Based on the distinct morphological and biochemical changes of the treated and dying cells, two modes of cell death have been described, namely apoptosis and necrosis. A number of anticancer drugs have been shown to induce apoptosis on varying cancer cell types. Apoptosis is a biochemically and genetically controlled response by which various types of cells can commit suicide in a programmed cell death fashion, and this effect has become a major focus for the study of cancer biology. Besides the morphology-based evaluation, flow cytometry assay has also been performed in this study to distinguish the mode of cell death induced by test herbal extracts. Results of flow cytometric analyses indicate that the levels of apoptotic peak (sub-G1 peak) were formed and increased after treating MCF-7 or other test cells with an herbal extract for 12 to 24 hours. A specific profile of programmed cell death was observed for MCF-7 and other test cells after the cells were treated with our target medicinal herbal extracts.

It has been proposed that two specific and separate signaling pathways are involved in the initiation of apoptosis in mammalian cell systems; one is mediated via the Fas/FasL pathway and another via the TNF-α pathway. In our experiments, some of the signaling molecules, including NF-κB, cyt c, p53, caspase 2, 3, 8 and others are found in our study to indeed involve in the apoptosis of our test cells, as were characterized by western-blot analysis or enzymatic activity assays. We report here that via a western-blotting analysis that the Rip protein has been apparently phosphorylated after the tumor cells were treated with a testing herbal extracts. Some very specific cellular events versus cell type activities have been obtained in our study, suggesting involvement of specific cyto-toxic biochemical mechanism.

5. METABOLITE PROFILING AND INDEX COMPOUNDS

Finally, we would like to address the question of approaches for need to standardize or normalize crude plant extracts and the derived fractionated compounds. With bioorganic chemistry, we are characterizing the index and/or bioactive compounds from our target herbal extracts. We have presently identified two new compounds, which have not been previously reported, and some other twenty compounds were also identified from our target herbs. Based on the structural features of these compounds, we have also evaluated their anti-oxidant activities. Results are being submitted for publications elsewhere. This project has been pursued as an integrated and fully coordinated works between the laboratories of Dr. L.-F. Shyur and N.-S. Yang. We hope that we can soon effectively employ these assay systems for evaluation of the old formulations and prescriptions of Chinese herbal medicines in terms of anti-tumor, cancer-prevention, and immune-modulation bioactivities.

REFERENCES

Ahonen, C. L., Gibson, S. J, Smith, R. M., Pederson, L. K., Lindh, J. M., Tomai, M. A., and Vasilakos, J. P. Dendritic cell maturation and subsequent enhanced T-cell stimulation induced with the novel synthetic immune response modifier R-848. Cellular Immunol. 197:62-72, 1999.

Banchereau, J and Steinman, R. M. Dendritic cell and control of immunity. Nature 392:245-252, 1998.

Banchereau, J., Briere, F., Caux, C., Davoust, J., Lebecque, S., Liu, Y.-J., Pulendran, B., and Palucka, K. Immunology of dendritic cells. Annu. Rev. Immunol. 18:767-811, 2000.

Chakraborty, A., Nitya, L. L., Chakraborty, G., and Mukherji, B. Stimulatory and inhibitory differentiation of human myeloid dendritic cells. Clin. Immunol. 94:88-98, 2000.

Cumberbatch, M. and Kimber, I. Dermal tumour necrosis factor-alpha induces dendritic cell migration to draining lymph node, and possibly provides one stimulus for Langerhans cell migration. Immunology 75:257-263, 1992.

De Smedt, T, Pajak, B., Muraile, E., et al. Regulation of dendritic cell number and maturation by lipopolysaccharide in vivo. J. Exp. Med. 184:1413-1424, 1996.

Förtsch, D., Röllinghoff, M., and Stenger, S. IL-10 converts human dendritic cells into macrophages-like cells with increased antibacterial activity against virulent Mycobacterium tuberculosis. J. Immunol. 165:978-987, 2000.

Kim, K. D., Lee, J. K., Park, S. N., Choe, I. S., Choe, Y.-K. Kim, S. J. Lee, E., and Lim, J.-S. Enhanced antigen-presenting activity and tumour necrosis factor-α-independent activation of dendritic cells following treatment with mycobacterium bovis bacillus Calmette-Guérin. Immunology 97:626-633, 1999.

Koide, S. L., Inaba, K., and Steinman R. M. Interleukin 1 enhanced T-cell-dependent immune responses by amplifying the function of dendritic cells. J. Exp. Med. 165:515-530

Larsson, M., Messmer, D., Somersan, S., Fonteneau, J.-F., Donahoe, S. M., Lee, M., Dunbar, P. R,Cerundolo, V., Julkunen, I., Nixon, D. F., and Bhardwaj, N. Requirement of mature dendritic cells for efficient activation of influenza A-specific memory CD8+ T cells. J, Immunol. 165:1182-1190, 2000.

McRae, B. L., Nagai, T., Semnani, R. T., van Seventer, J. M., and van Seventer, G. A. Interferon-α and -β inhibit the in vitro differentiation of immunocompetent human dendritic cells from CD14+ precursors. Blood 96:210-217, 2000.

Piemonti, L. , Monti, P., Allavena, P., Sironi, M., Soldini, L, Leone, B. E., Socci, C., and Di Carlo, V. Glucocorticoids affect human dendritic cell differentiation and maturation. J. Immunol. 162:6473-6481, 1999.

Piemonti, L. , Monti, P., Sironi, M., Fraticelli, P., Leone, B. E., Cin, E. D., Allavena, P., and Di Carlo, V. Vitamin D3 affects differentiation, maturation, and function of human monocyte-derived dendritic cell. J. Immunol. 164:4443-4451, 2000.

Pulendran, B., Smith, J.L., Jenkins, M., Schoenborn, M., Maraskovsky, E., and Maliszewski, C. R. Prevention of peripheral tolerance by dendritic cell growth factor: Flt3 ligand as an adjuvant. J. Exp. Med. 188:2075-2082, 1998.

Sallusto, F. and Lanzavecchia, A. Efficient presentation of soluble antigen by cultured human dendritic cells is maintained by granulocytes/macrophages colony-stimulating factors plus interleukin-4 and down regulated by tumor necrosis factor alpha. J. Exp. Med. 179:1109-1118, 1994.

Chapter 9

USING TRANSCRIPTION FACTOR BASED ASSAYS TO STUDY HERBAL PRODUCTS

DAVID S. PASCO
National Center for the Development of Natural ProductsUniversity of Mississippi University, MS 38677

Abstract:
Transcription factors bind to specific DNA sequences in the vicinity of genes and are the major regulators of gene expression. The activity of these factors is controlled by the various signal transduction mechanisms and second messenger systems used by the cell to coordinate and respond to signals from the extracellular environment, including hormones and the local tissue environment. Because of this, transcription factors can be used to indirectly monitor the activity of many cellular components, such as the status of membrane receptor-hormone interactions, the biochemical transformations accompanying the relay of this binding event to a cytoplasmic signal, or the multitude of protein targets used to coordinate this signal to various cellular compartments. By monitoring the activity of several transcription factors, one can gain an overall impression of the state of various signal transduction systems and other targets. Therefore, monitoring transcription factors' activities is useful for determining the influence that a particular herb or herbal component may have on a cell.
The easiest way to detect changes in transcription factor activity is through luciferase reporter gene technology. These plasmid constructs can contain either the complex promoter region of a gene or binding sites for a particular transcription factor. Comparing activity profiles of herbs or herbal components to specific transcription factors can provide an index of specificity, in addition to determining potential mechanisms of action. The particular transcription factor binding sites comprising the battery are chosen based on the type of activity an herb exhibits and/or upon a well-established therapeutic target of the disorder. The same criteria determine the cell type or types used for the assays. Results and examples of the utilization of this technology over the past several years at the National Center for the Development of Natural Products (NCNPR) will be presented from the therapeutic areas of inflammation and immunomodulation.

Yuan Lin (ed.), Drug Discovery and Traditional Chinese Medicine: Science, Regulatory and Globalization, 83-87.
©2001 *Kluwer Academic Publishers. Printed in the Netherlands.*

1. ANTIINFLAMMATORY ASSAY

Recently, anti-inflammatory strategies have focused on the development of specific inhibitors of the inducible form of cyclo-oxygenase (COX), COX-2. An equally plausible approach could employ compounds that suppress the synthesis of COX-2. The NCNPR has developed a high throughput luciferase reporter gene assay to identify novel inhibitors of COX-2 gene expression that utilizes the promoter of this gene to drive luciferase expression. Secondary assays using a battery of luciferase reporter constructs, containing the binding sites for transcription factors found within the COX-2 gene in addition to Sp1 binding sites, were useful for further characterizing actives and identifying extracts or compounds that were cytotoxic or that inhibited some aspect of the luciferase assay (Figure 1).

Extracts from plants used traditionally for inflammation were tested in this assay and several, such as *Curcuma longa* (turmeric) and *Arnebia euchroma*, were very active. *A.euchroma* extract is used in a topical, oil based ointment to treat burns, skin rashes and fungal infections and has been used in TCM for centuries. In addition to inhibiting induction of the COX-2 promoter, this extract also suppressed the activation of the transcription factors AP-1 and NF-kappa (Figure 1). Two factors important for coordinating the expression of a multitude of genes expressed during the inflammatory response. Shikonins, a class of compounds found within *A. euchroma*, were previously found to be responsible for anti-inflammatory activity in animal studies. Shikonin, in addition to five naturally occurring shikonin analogs, exhibited the same activity in the reporter gene assay as the crude extract. These compounds also inhibited the induction of the COX-2 protein and mRNA in cells treated with inflammatory agents.

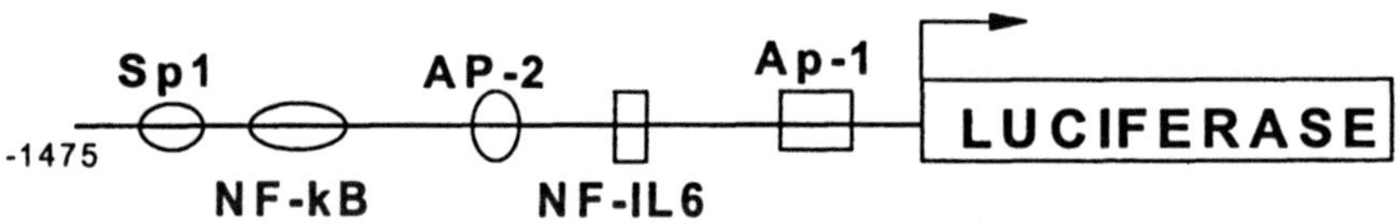

Fig. 1. Anti-inflammatory assay with COX-2 promoter

Additional studies in this area showed that out of 32 plants used traditionally for inflammation, extracts from five were potent inhibitors of NF-kappaB activation and 10 displayed inhibition of COX-2 enzyme activity. In addition, of the active plant extracts two of these were active in both assays. These studies show that medicinal plants can not only impact inflammation via the inhibition of the COX-2 enzyme but also by inhibiting the activation of transcription factors that regulate the expression of genes during inflammation.

2. IMMUNE STIMULATION

The macrophage, in working with B- and T-lymphocytes, induces responses to antigens and invading microorganisms, develops potent antimicrobial cytotoxic mechanisms, and mediates the destruction of host tissue including transformed cells. To detect potential immunostimulatory activities within herbal products we employed a transcription factor-based bioassay for NF-kappaB in THP-1 human monocytes/macrophages (Figure 2). NF-kappaB mediates many of the signals that characterize the activated state in this cell type.

Fig. 2: Macrophage Activation

Aqueous extracts from several botanicals were quite potent activators of NF-kappaB in these cells and included those from aloe vera and three micro algae used commercially as dietary supplements.

3. ALOE VERA POLYSACCHARIDE

The gel from the *Aloe Vera plant* has been used for centuries to treat burns and other wounds to increase the rate of healing. One mechanism by which this gel could influence wound healing is by the activation of macrophages. Acemannan, a large molecular weight polydispersed beta- (1,4)-linked acetylated mannan interspersed with O-acetyl groups, comprises approximately 5% of the dry weight of this gel. *In vitro* experiments have shown that acemannan can activate macrophages, but the degree of activation is very low and the concentrations of acemannan required were very high (200 to 1000 µg/ml). This led to a hypothesis that either acemannan was not very potent or that trace amounts of a potent compound was present as a "contaminant" within acemannan preparations.

Preliminary findings from the NCPNR laboratory suggest that the level of activity for macrophage activation in crude aloe juice and gel could not be accounted for by known aloe components such as acemannan. Initial fractionation of crude aloe juice demonstrated that the major portion of this activity was present in the fraction above 500,000 Daltons. This fraction maximally activated macrophages at a concentration of 50 micrograms/ml. This fraction was further fractionated using HPLC size exclusion chromatography and the chromatogram showed two major peaks, one with a molecular weight of between 4 and 7 million Daltons (peak A) and the other between 0.1 and 1.0 million (peak B). Peak A was at least 10 times more active than peak B in the macrophage NF-kappaB assay. Two preparations of acemannan (AVMP and Manapol) were analyzed using this HPLC system, and the acemannan peaks had the same retention times as peak B. Traces of a peak identical in retention time to peak A were also evident in this chromatograph. Peak A is thus chromatographically distinct from acemannan and is at least 10,000 times more potent than this polysaccharide for activating macrophages. Chemical analysis of purified peak A indicated that it is 80-100% carbohydrate and is composed mainly of glucose, galactose, mannose, and arabinose. On a per weight basis, it is as active as *E. coli* lipopolysaccharide in the NF-kappaB macrophage activation assay (EC_{50}=0.5 micrograms/ml), and no detectable levels of endotoxin were present in the preparation, as evidenced by a lack of beta-hydroxymyristate in the glycosyl composition analysis. The polysaccharide in peak A also induced the expression of mRNAs for IL-1 beta and TNF-alpha to levels equal to lipopolysaccharide confirming macrophage activation. Although this polysaccharide comprised only 0.015% of the original dry weight of aloe juice, its potency for macrophage activation accounts fully for the immunostimulatory activity of the crude juice.

4. MICRO ALGAE POLYSACCHARIDES

Several types of microalgaes are used as nutrient-dense food sources and are common dietary supplements sold internationally. Anecdotal reports and several studies demonstrate that consumption of these microalgae may have beneficial health effects, including enhancement of immune function. Initial testing of aqueous extracts of three microalgae, *Aphanizomenon flos-aquae, Spirulina plantensis, and Chlorella pyrenoidosa.* in the macrophage NF-kappaB assay indicated extremely high levels of activity. Following solvent partitioning, alcohol precipitation, and ultrafiltration, the activities were found to be in the fraction above 100,000 Daltons. Size exclusion HPLC chromatography of this high molecular weight fraction indicated that for all three algae the activity was confined to one peak with a size between 5 and 10 million Daltons. The carbohydrate content of each isolated microalgal polysaccharide was estimated to be between 90% and 100% with activities in the NF-kappaB assay ranging from EC_{50} of 20 to 110 ng/ml. Endotoxin was not present in two of the polysaccharide preparations while it was present in the *A. flos-aquae* polysaccharide isolates at extremely low levels (0.6% of total peak area). Treatment of macrophages with these polysaccharides in the presence of polymyoxin B did not attenuate the response, also indicating that these activities were not due to endotoxin. Glycosyl composition analysis for each polysaccharide indicated that each was unique. Macrophage activation was confirmed by experiments demonstrating the induction of IL-1 beta and

TNF-alpha mRNAs to levels equal to those found in cells maximally stimulated with bacterial lipopolysaccharide. The development of these microalgal polysaccharides could potentially add to the arsenal of available agents for immunotherapy in the treatment of cancer and infectious diseases.

ACKNOWLEDGEMENTS

Dr. Andrew Dannenberg and Dr. Predrag Bulic were collaborators on the anti-inflammatory assays. Dr. Yuan Lin was a collaborator on the shikonin isolation. Nirmal Pugh was a collaborator with the aloe and microalgae work, and Dr. Samir Ross and Dr. Mahmoud ElSohly were collaborators with the polysaccharide work.

Chapter 10

MOLECULAR BASIS FOR MEDICINAL ACTIONS OF ANDROGENS AND GREEN TEA EPIGALLOCATECHIN GALLATE

SHUTSUNG LIAO, YUNG-HSI KAO, MAI T. DANG, CHING SONG, JUNICHI FUKUCHI, JOHN M. KOKONTIS, AND RICHARD A. HIIPAKKA
Tang Center for Herbal Medicine Research, Ben May Institute for Cancer Research and Department of Biochemistry and Molecular Biology, University of Chicago, Chicago, IL 60637

Abstract: Androgens, which have been around for more than 2000 years, are probably the oldest drugs used in purified forms, while green tea extracts, which have been around for more than 3000 years, are among the most widely used ancient medicinal agents in man's history. It is now clear that a specific green tea catechin, (-)epigallocatechin-3-gallate (EGCG), can regulate biological actions of androgens and other hormones. Androgens and EGCG can also modulate the growth of androgen-dependent and independent prostate cancers, as well as other hormone related abnormalities.

Keywords: Testosterone, catechin, 5α-reductase, benign prostatic hyperplasia, prostate cancer, acne, cell cycle arrest, apoptosis, food intake restriction, obesity.

1. HISTORICAL INTRODUCTION

As early as 200 BC, the Chinese were using androgenic preparations to treat individuals lacking 'maleness activity'. According to Chinese history, emperor Shen-Nong found, more than 3,000 years ago, that a daily cup of tea could dissolve many poisons in the body. In oriental culture, it has been widely believed for a long time that tea has medicinal efficacy for prevention and treatment of many diseases. Joseph Needham, a distinguished biochemist

Yuan Lin (ed.), Drug Discovery and Traditional Chinese Medicine: Science, Regulatory and Globalization, 89-96.
©2001 *Kluwer Academic Publishers. Printed in the Netherlands.*

in England, stated that this was the most important pre-biochemical discovery. Scientific and medical evaluation of tea, however, started only very recently.

Recent studies indicate that specific green tea catechins can modulate androgen actions, *in vivo*, and therefore, can be medically important for treating male hormone-related abnormalities.

2. CONTROL OF ANDROGEN ACTION

In the early 1960s, it was found that androgens can rapidly enhance RNA synthesis in target organs, such as the ventral prostate of rats, suggesting that androgens act by modulating gene expression. Subsequent studies have shown that, in the target organs, testosterone, the major androgen produced by testis and circulating in blood, is converted by 5α-reductase to 5α-dihydrosterone (DHT) that binds to a specific nuclear receptor (AR). The DHT-AR complex, in conjunction with other chromosomal proteins, then regulates the synthesis of specific RNA and modulates cellular activities and organ functions. Mutations in the genes for 5α-reductase or AR have been shown to be responsible for androgen-insensitivity syndromes.

The molecular steps required for androgen action provide two effective methods for control of testosterone-regulated responses: (a) the use of a 5α-reductase inhibitor to suppress DHT production and (b) the use of antiandrogens to block the interaction of DHT with AR. Both methods are now being utilized as therapies for androgen-related disorders.

3. EGCG AND OTHER NATURAL
5α-REDUCTASE INHIBITORS

Two isozymes of 5α-reductase have been identified, but the specific roles of the individual isozymes are not well understood. The synthetic 4-aza-steroid, finasteride, prescribed for benign prostate hyperplasia, is a selective inhibitor of the type 2 isozyme of 5α-reductase and covalently reacts at the enzyme active site with NADP to form a potent bisubstrate inhibitor.

A number of natural compounds that inhibit 5α-reductase have been found. The first one we identified in 1992 was γ-linolenic acid [C18:3 (cis-9,12,15)], an essential fatty acid in many plant oils, including evening primrose and borage oil that are being used as health food products. In cell-free assay systems, free fatty acids with carbon chain length of more than 14 and one or more double bonds are active at 40 μM. Conjugated fatty acids (glycerides or esters) are not active at 100 μM. γ-Linolenic acid is far more active than many dozens of other fatty acids tested and is active at concentrations lower than 5 μM.

Subsequently, we found that certain green tea catechins were also active 5α-reductase inhibitors. Catechin gallates, such as EGCG, (-)epicatechin-3-gallate (ECG), and (-)catechin-3-gallate, are active at concentrations less than 5 μM. The gallate group is important for inhibitory activity. Non-gallated catechins, such as (-)epicatechin (EC) or (-)epigallocatechin (EGC) are not active. Gallic acid and the methyl ester of gallic acid are also not active.

The biological activity of γ-linolenic acid and EGCG has been tested *in vivo* using flank organs of male hamsters as animal models. When hamsters are castrated, flank organ growth is clearly suppressed. Topical application of testosterone on the flank organ stimulates the growth of the organ, but this growth is effectively suppressed by topical application of γ-linolenic acid or EGCG. Topical application of γ-linolenic acid or EGCG to the forehead of a man also inhibited sebum production.

γ-Linolenic acid can inhibit both the type 1 and type 2 isozymes, whereas EGCG and ECG inhibited only type 1 isozyme of 5α-reductase at concentrations less than 30 μM. The biomedical utility of this difference is not clear at this time. The ability of γ-linolenic acid and EGCG to interfere, *in vivo*, with DHT-dependent sebum production and flank organ growth suggests that these natural substances, as well as other natural 5α-reductase inhibitors, may be helpful in the treatment of other skin problems including acne and baldness.

4. SUPPRESSION OF PROSTATE TUMORS BY ANDROGEN

Since prostate cancer is initially androgen-dependent, hormonal therapy, pioneered by Charles Huggins 60 years ago using castration or more recently antiandrogens, has been the front-line therapy for prostate cancer. More than 70% of patients benefit from this therapy. However, prostate cancer recurs in most of these patients in one to three years as tumors that do not need androgen for growth. For lack of effective therapy, patients die from this androgen-independent cancer.

During the last 10 years, we established that androgen-independent cancer cells could develop from clonal androgen-dependent cancer cells during long-term (1-3 years) culture in androgen-depleted media. This transition to androgen independence is accompanied by dramatically increased AR expression without gene amplification or new mutation in the AR gene. Surprisingly, the growth of these cells is inhibited by physiological levels of androgens.

These androgen-independent prostate cancer cells grow as tumors in castrated athymic mice but not in normal athymic male mice. Administration of androgen to castrated mice prevents tumor growth and suppresses prostate tumors already present in these animals.

The 5α-reductase inhibitor, finasteride (Proscar) or antiandrogens, such as Casodex, block the repressive effect of testosterone on these xenografts and stimulated the tumor growth, suggesting that the growth suppression requires conversion of testosterone to DHT and binding of DHT to AR. If testosterone has a role in suppressing prostate cancer growth in patients, the use of these drugs may enhance the growth of certain prostate cancers.

The DHT-dependent suppression of prostate cancer cell proliferation appears to be dependent on cell cycle arrest due to increased levels of Cdk2 inhibitors including p27.

5. SUPPRESSION OF PROSTATE AND BREAST TUMORS BY EGCG

Green tea consumption has been linked to lower incidence of cancer of the stomach and certain organs, and numerous studies have shown tumor suppression by lengthy use of green tea beverage in animals. However, many epidemiological studies have not provided strong evidence that green tea is clearly antitumorigenic in humans.

For better understanding of the ability of green tea to control cancer growth, we injected purified catechins intraperitoneally (ip) into tumor-bearing mice. Human prostate cancer cell lines, PC-3 (AR negative and androgen insensitive) and LNCaP 104-R (androgen-repressive) inoculated subcutaneously into nude mice produce prostate tumors. Green tea EGCG, ip injected, significantly inhibits the growth and rapidly (in a week) reduces the size of human prostate tumors in athymic mice. Structurally related catechins, such as ECG that lack only one of the eight-hydroxyl groups in EGCG are totally inactive. EC and EGC are also not antitumorigenic.

Both androgen-dependent and androgen-independent prostate tumors respond to tumor suppression by EGCG, suggesting that EGCG action was not related to modulation of androgen activity. In addition, the growth of human breast tumors in nude mice produced by human breast cancer, MCF-7 cells are also clearly inhibited during the first week of ip injection of EGCG.

It is possible that the low clinical incidence of prostate and breast cancer in some Asian countries is related to high green tea consumption. The frequency of the latent, localized prostate cancer does not vary significantly among geographically different populations, but the clinical incidence of metastatic prostate cancer varies considerably among countries (low in Japan and high in the United States). If consumption of green tea beverage is related to this difference, EGCG may play an important role in preventing the progression or metastasis of prostate cancer cells.

6. MODULATION OF FOOD INTAKE AND OBESITY BY EGCG

In oriental countries, long-term use of green tea beverage is considered to be beneficial for keeping a healthy body weight. However, clear scientific evidence has not been available until recently. EGCG, given to rats by ip injection, could within 2 to 7 days reduce body weight by about 20 to 30%. Other structurally related catechins, such as EC, EGC, or ECG, are not effective at the same dose. The body weight loss is reversible; when EGCG administration is stopped, animals regain body weight. Reduction of body weight appears to be due to EGCG-induced reduction in food intake. The loss of appetite might involve neuropeptide(s) other than leptin, since EGCG is effective in reducing body weight of lean and obese (leptin-receptor negative) female and male rats.

Various hormones including cholecystokinin, glucagon-like peptide-1, glucagon, substance P, somatostatin, and bombesin, have been reported to inhibit food intake, and it has been reported that plasma cholecystokinin levels are elevated in rats given a diet supplemented with tea polyphenols. Further study is required to determine whether the expressions of other hypothalamic or gastrointestinal neuropeptide genes that control appetite are altered by EGCG and perhaps responsible for the effect of EGCG on food intake.

Although orally administered EGCG is not as effective as ip injected EGCG, probably due to poor absorption of EGCG from the intestine, long-term oral use of green tea beverage (2-4 cups/day) or EGCG-containing drinks may mimic the effects of ip injected EGCG. Since EGCG can also selectively reduce body fat accumulation, EGCG may be useful for treatment of obesity.

Using a cell culture system, we have found that EGCG reduces total triglyceride accumulation in murine 3T3-L1 preadipocytes during their differentiation into adipocytes or in differentiated 3T3-L1 adipocytes.

7. MODULATION OF ENDOCRINE SYSTEMS BY EGCG

Although many health benefits of tea consumption have been demonstrated, the effects of tea on endocrine systems have not been evaluated very carefully until recently. It is now known that rats ip injected with EGCG have significant changes in various endocrine parameters. After seven days of ip treatment with EGCG, circulating levels of testosterone are reduced by about 75% in male rats and 17β-estradiol levels by 34% in female rats. The weight of androgen-sensitive organs, such as ventral prostate, seminal vesicles, coagulating and preputial glands are reduced by 50 to 70% after EGCG treatment. Similarly, the weight of estrogen-sensitive organs, such as the uterus and ovary of female rats is reduced by about 50% after EGCG treatment. These changes in the weight of sexual organs are catechin-specific, with EGCG showing the largest effect. The effect of EGCG on the prostate or uterine weight loss is due to reduced sex hormone levels and not a direct effect of EGCG on the organs; the organ weight loss is completely reversed with externally supplied sex hormones. With male and female rats treated with EGCG for seven days, the serum level of LH are significantly reduced by 40 to 50%, suggesting that low LH production led to the reduced blood levels of sex hormones.

In both male and female rats, seven days of EGCG treatment causes significant reduction in blood levels of leptin, IGF-1 and insulin. The effect of EGCG on these peptide hormones is not mimicked by structurally similar catechins, EC, EGC, or ECG at an equivalent dose.

In EGCG-treated male rats, the serum level of protein, fatty acids and glycerol are not altered, but significant reductions in serum glucose (-32%), lipids (-15%), triglycerides (-46%) and cholesterol (-20%) are observed. Based on proximate composition analysis, rats treated daily with EGCG for seven days have no change in percent water and protein content, a moderate decrease in carbohydrate content, and a very large reduction in fat content, decreasing from 4.1% in control to 1.4% in EGCG-treated group. Within seven to eight days, EGCG treatment decreases subcutaneous fat by 40 to 70% and abdominal fat by 20 to 35% in male rats.

8. ROUTE OF EGCG ADMINISTRATION AND HEALTH BENEFITS

The effects of EGCG on body weight loss, hormone level changes, and food intake depend on the route of administration. The effects of EGCG observed when EGCG is administered by ip injection are not present when the same amount of EGCG is given to rats orally. This may be due to inefficient absorption of EGCG or metabolism in the digestive tract and suggest that the effects of ip administration of EGCG are not caused by interaction of EGCG with food or by EGCG action inside the gastrointestinal tract.

Although oral administration of EGCG is less effective, long-term oral consumption of green tea or EGCG-containing extracts may mimic some of the acute EGCG effects caused by ip administration of EGCG and may be beneficial to health. Based on oral and ip effects of EGCG on serum hormones and nutrients, long-term consumption of green tea may influence the incidence and provide therapies for various diseases.

By lowering plasma levels of sex steroids and other endocrine factors, such as IGF-1, long-term use of EGCG or green tea may be effective in prevention and suppression of the growth of hormone-dependent and hormone-independent cancers of various organs.

9. CONCLUDING REMARKS

Androgen may be the first natural medicine used in purified form for therapy. It provides the best example that traditional Chinese Medicines are not necessarily effective only in combination. Unfortunately, many Chinese medicine researchers and practitioners are not aware of this great discovery and still support strongly an anti-molecular approach. It is the time to correct this misconception.

Green tea beverage originated many thousands of years ago as a medicinal tonic. Although the historically long use of many folk remedies does not necessary prove their medical usefulness, recent scientific evidence, appears to support the possibility that green tea catechins are medically valuable. However, it is important to consider also the potential adverse effects that may accompany the use of green tea or catechins. For example, alteration of endocrine systems may have serious consequence in pregnant woman and small children.

REFERENCES

This article is based mainly on the studies carried out in the authors' laboratory. Major studies cited in this article can be found in the following publications.

Hiipakka, R. A., and Liao, S. (1998). Molecular mechanism of androgen action. *Trends Endocrinol. Metab.* **9,** 317-324.

Kao, Y. H., Hiipakka, R. A., and Liao, S. (2000). Modulation of endocrine systems and food intake by green tea epigallocatechin gallate. *Endocrinology* **141,** 980-987.

Kao, Y. H., Hiipakka, R. A., and Liao, S. (2000). Modulation of obesity by a green tea catechin. *Am. J. Clin. Nutr.* In Press.

Kokontis, J. M., Takakura, K., Hay, N., and Liao, S. (1994) " Increased androgen receptor activity and altered c-myc expression in prostate cancer cells after long-term androgen deprivation" *Cancer Research* **54,** 1566-1573..

Kokontis, J. M., Hay, N., and Liao, S. (1998) "Progression of LNCaP prostate tumor cells during androgen deprivation: Hormone-independent growth, repression of proliferation by androgen, and role for p27^{kip1} in androgen-induced cell cycle arrest" *Mol. Endocrinol.* **12,** 941-953.

Kokontis, J. M., and Liao, S. (1999). Molecular action of androgen in the normal and neoplastic prostate. *Vitam. Horm.* **55,** 219-307.

Liang, T., and Liao, S. (1992). Inhibition of 5α-reductase by specific aliphatic unsaturated fatty acids. *Biochem. J.* **285,** 557-562.

Liang, T., and Liao, S. (1997). Growth suppression of hamster flank organ organs by topical application of γ-linolenic and other fatty acid inhibitors of 5α-reductase. *J. Invest. Dermatol.* **109,** 152-157.

Liao, S., and Fang, S. (1969) "Receptor proteins for androgens and the mode of action of androgens on gene transcription in ventral prostate". *Vitamins and Hormones* **27,** 17-90.

Liao, S., and Hiipakka, R. A. (1995). Selective inhibition of steroid 5α-reductase isoenzymes by tea epicatechin-3-gallate and epigallocatechin-3-gallate. *Biochem. Biophys. Res. Commun.*

Liao, S., Kokontis, J., Sai, T., and Hiipakka, R. A. (1989) "Androgen receptors: Structures, mutations, antibodies and cellular dynamics". *J. Steroid Biochem.* **34**, 41-51.

Liao, S., Umekita, Y., Guo, J., Kokontis, J. M., and Hiipakka, R. A. (1995). Growth inhibition and regression of human prostate and breast tumors in athymic mice by tea epigallocatechin gallate" *Cancer Letter_* **96**, 239-243.

Liao, S., Kao, Y. H., and Hiipakka, R. A. (2000) "Green tea: Biochemical and biological basis for health benefits" *Vitamins and Hormones* (in press).

Umekita, Y., Hiipakka, R. A., Kokontis, J. M., and Liao, S. (1996) "Human prostate tumor growth in athymic mice: Inhibition by androgens and stimulation by finasteride" *Proc. Natl. Acad. Sci. U. S. A.,* **93,** 11802-11807.

Chapter 11

STUDIES ON CHEMICAL COMPONENTS AND THEIR PHARMACOLOGICAL ACTIVITIES OF *PANAX GINSENG* ROOT

LI-XIANGAO, LI XIANG AND LEI JUN
College of Chinese Medicine Materials, Jilin Agricultural University, China 130118

Abstract: This paper explains the details of how ginsenosides are isolated and extracted from the roots of *Panax ginseng*. Fresh ginseng homogenate and red ginseng powder were soaked with MeOH, which resulted in a concentrated extract. The extract was then diluted with water and Et$_2$O was added. The water phase was further extracted by water-saturated n-BuOH, and the new extract was chromatography on Macro-reticular resin column. Water was used as elutriation to remove impurities, such as sugar and water-soluble substances, and the extract was further eluted with a different fraction of MeOH. Every part was chromatography on silica gel column, and the individual ginsenosides were isolated. The identity of these ginsenosides was obtained after hydrolysis, partial hydrolysis, methylation, permethylation, methanolysis, and acethylation. Various types and structures of the ginsenosides are included in this report. Finally, each ginsenoside was tested for relevant pharmacological action. The purpose of this paper is to provide a molecular basis for the many health benefits received by taking ginseng.

1. INTRODUCTION

"Renshen", the dry root of Panax ginseng C. A. Mey (Araliaceae), is the most widely used Traditional Chinese Medicine in the world. Wild-growing or cultivated ginseng root is officially listed in the Chinese Pharmacopoeia and used as a tonic.

Despite its long history of usage, the isolation and characterization of the chemical constituents of ginseng did not occur until the 1960's. This report is intended to provide the molecular basis for the health benefit claims and pharmacological actions recorded in the Chinese Pharmacopoeia. The results of the experiments on ginseng also reveal new biological activities, such as preventing proliferation of cancer cells, controlling metastasis of carcinoma, and having the activity of dissolving fiber and preventing thrombosis.

Yuan Lin (ed.), Drug Discovery and Traditional Chinese Medicine: Science, Regulatory and Globalization, 97-109
©2001 *Kluwer Academic Publishers. Printed in the Netherlands*

2. ISOLATION OF GINSENOSIDES FROM ROOTS OF *PANAX GINSENG*:

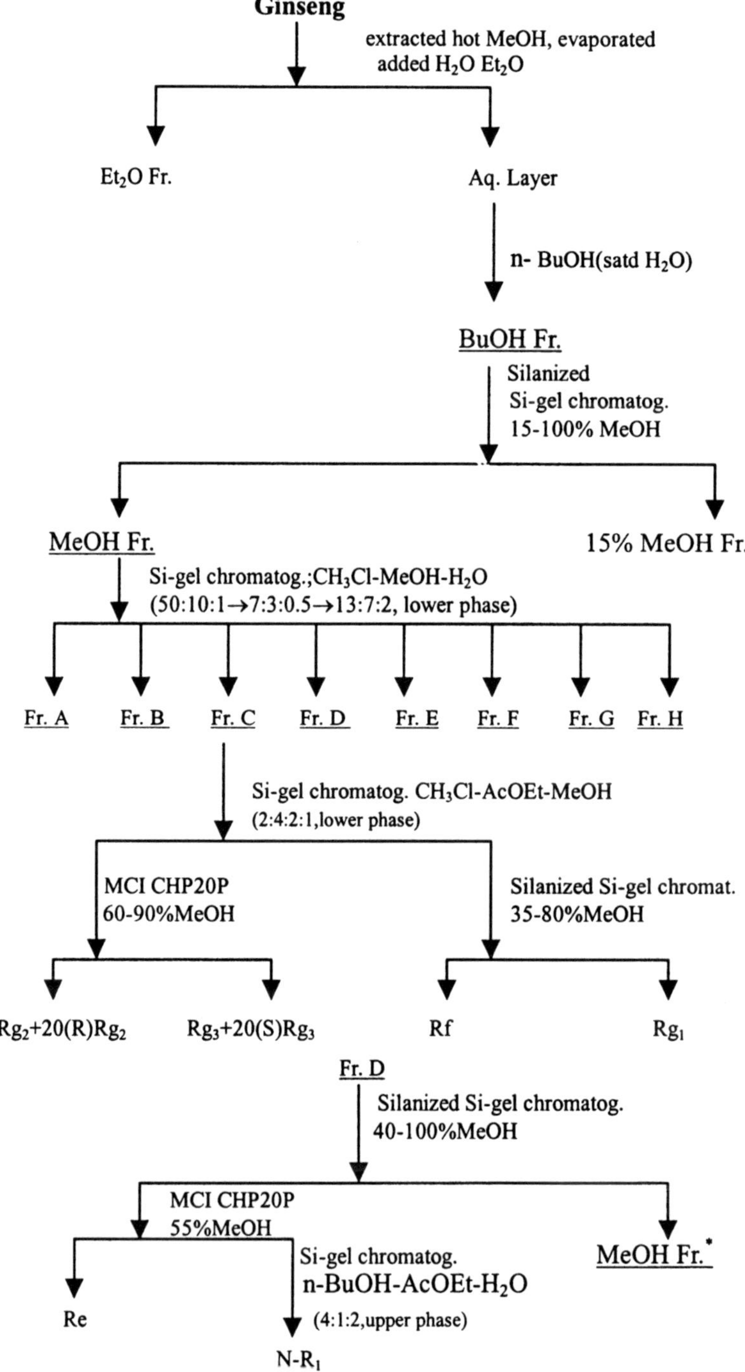

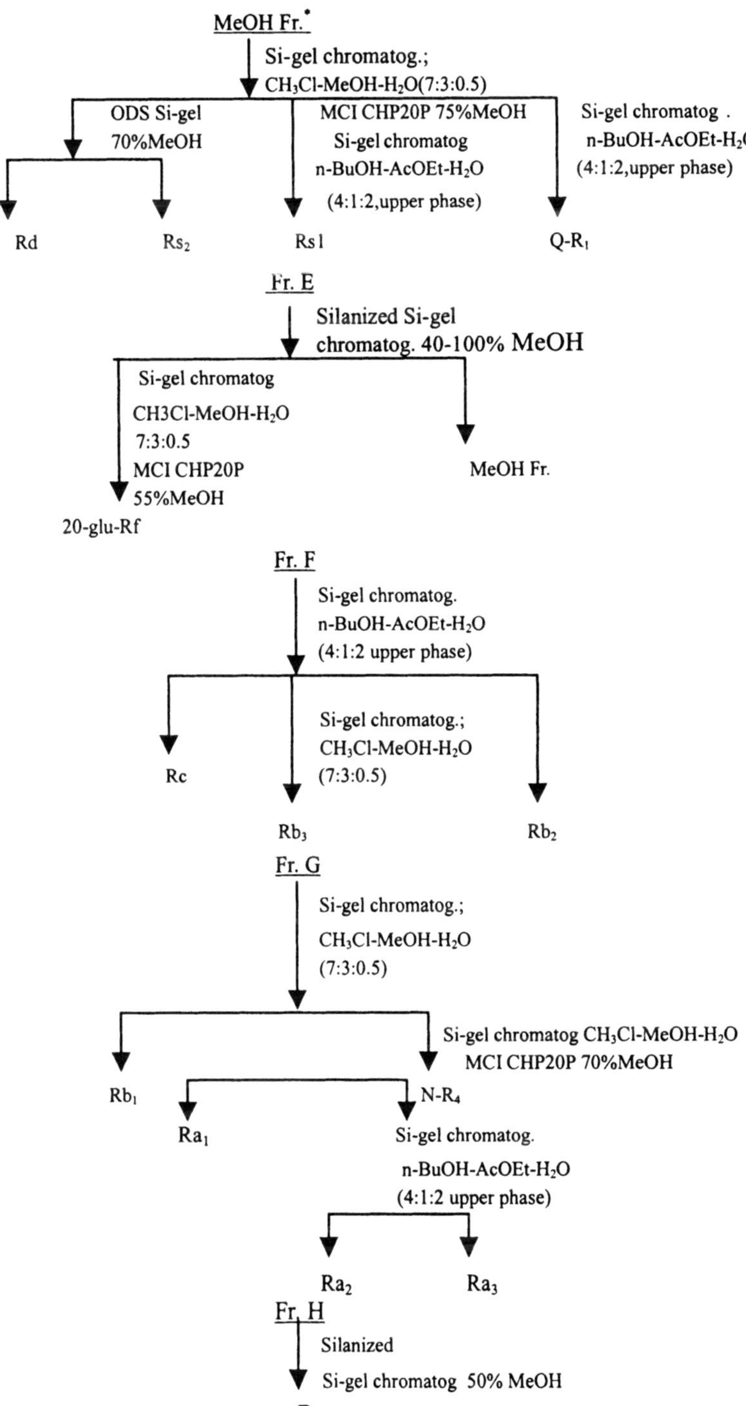

Figure1: Extraction and isolation flow of ginsenoside (Li Xianggao)

3. TYPES AND STRUCTURES OF GINSENOSIDES

3.1 CHEMICAL STRUCTURE OF PROTOPANAXADIOL SAPONIN

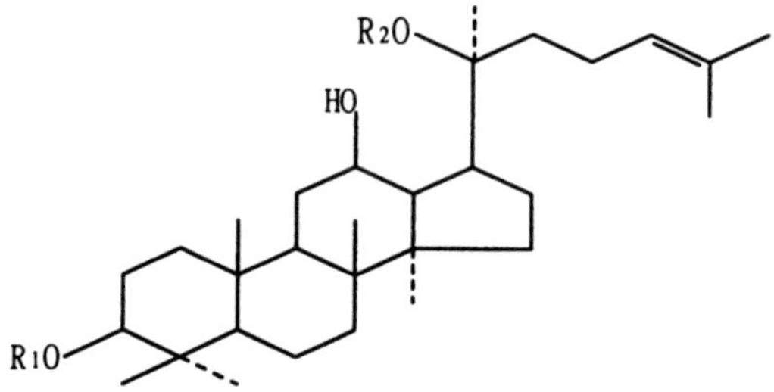

Figure 2: The chemical structure of protopanaxadiol saponin

Table 1: Chemical constituents in protopanaxadiol saponin.

	R_1	R_2
Ginsenoside-Ra$_1$	-glc(2,1)glc	-glc(6,1) arap(4,1)x
Ginsenoside-Ra$_2$	-glc(2,1)glc	-glc(6,1) arap(2,1)xy
Ginsenoside-Ra$_3$	-glc(2,1)glc	-glc(6,1) arap(3,1)x
Ginsenoside-Rb$_1$	-glc(2,1)glc	-glc(6,1) glc
Ginsenoside-Rb$_2$	-glc(2,1)glc	-glc(6,1) arap
Ginsenoside-Rb$_3$	-glc(2,1)glc	-glc(6,1) xyl
Ginsenoside-Rc	-glc(2,1)glc	-glc(6,1) araf
Ginsenoside-Rd	-glc(2,1)glc	-glc
Ginsenoside-Rg$_3$	-glc(2,1)glc	-H
Ginsenoside-F$_2$	-glc	-glc
Ginsenoside-Rh$_2$	-glc	-H
Quinquenoside-R$_1$	-glc(2,1)glc(6)Ac	-glc(6,1) glc
Ginsenoside-Rs$_1$	-glc(2,1)glc(6)Ac	-glc(6,1) arap
Ginsenoside-Rs$_2$	-glc(2,1)glc(6)Ac	-glc(6,1) araf
Malonyl-Ginsenoside-Rb$_1$	-glc(2,1)glc(6)Ma	-glc(6,1) glc
Malonyl-Ginsenoside-Rb$_2$	-glc(2,1)glc(6)Ma	-glc(6,1) arap
Malonyl-Ginsenoside-Rc	-glc(2,1)glc(6)Ma	-glc(6,1) araf
Malonyl-Ginsenoside-Rd	-glc(2,1)glc(6)Ma	-glc
Notoginsenoside-R4	-glc(2,1)glc	-glc(6,1) glc(6,1)xyl
Notoginsenoside-Fa	-glc(2,1) glc(2,1)xyl	-glc(6,1) glc
Gypenoside-XVII	-glc	-glc(6,1) glc

3.2 CHEMICAL STRUCTURE OF PROTOPANAXATRIOL SAPONIN

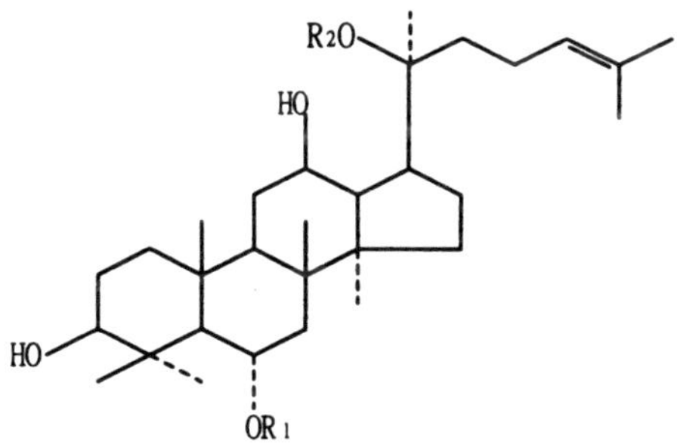

Figure 3: The chemical structure of protopanaxatriol saponin

Table 2: Chemical constituents of protopanaxatriol saponin.

	R_1	R_2
Ginsenoside-Re	Glc-Rha-	Glc
Ginsenoside –Rf	Glc-glc-	H
Ginsenoside –Rg$_1$	Glc-	Glc
Ginsenoside –Rg$_2$	Glc-Rha-	H
Ginsenoside-Rh$_1$	Glc-	H
20-gluco-ginsenoside-Rf	Glc-glc-	Glc

3.3 THE STRUCTURE OF GINSENOSIDES WITH TWO DOUBLE BONDS

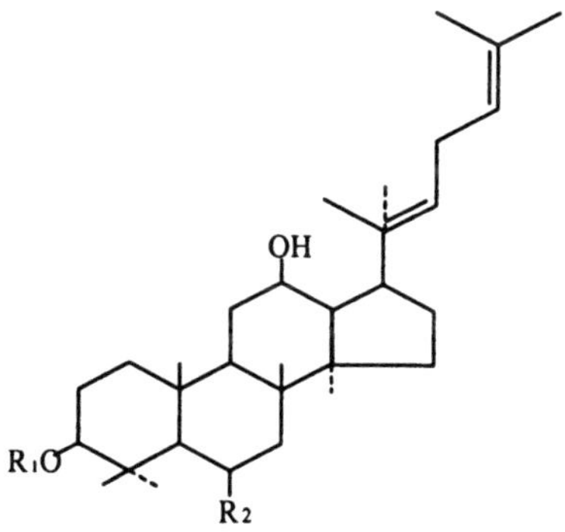

Figure 4: The chemical structure of ginsenosides with two double bonds

Table 3: The chemical constituents of ginsenosides with two double bonds.

	R_1	R_2
Ginsenoside-F$_4$	H	Rha-glc-O-
Ginsenoside –Rh$_3$	Glc-	H
Ginsenoside –Rg$_5$	Glc-glc	H
Ginsenoside –Rh$_4$	H	Glc-O-

3.4 THE STRUCTURE OLEANOLIC ACID TYPE GINSENOSIDES

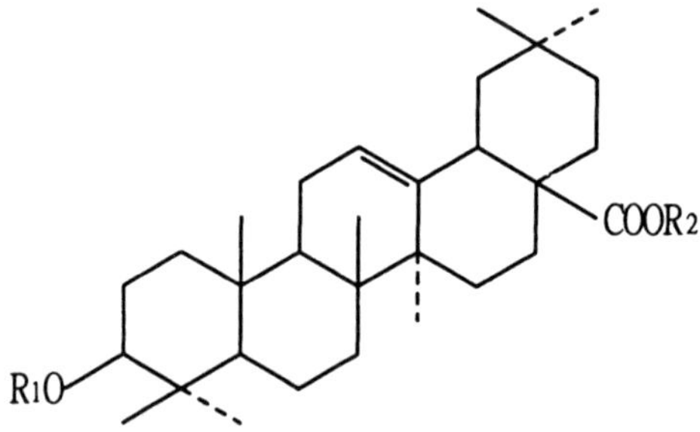

Figure 5: The chemical structure of oleanolic acid type ginsenosides

Table 4: The chemical constituents of Oleanolic acid type ginsenosides.

	R_1	R_2
Ginsenoside-Ro	GlcUA(2-1)-Glc	Glc-

Note: GlcUA is glucuronic acid

4. PHAMACOLOGICAL ACTIVITY OF GISENOSIDES

4.1 GINSENOSIDE-RB$_1$, RG$_1$ ACTIVITY ON NERVE CELL

4.1.1 EFFECT OF GINSENOSIDE-RB$_1$, RG$_1$ ON THE GROWTH OF MOUSE BRAIN NERVE

A batch of 50 ablactation mice of Kunming kind were randomly divided into five groups. The mice were given water with ginsenoside-Rb1, Rg1 0.125 mg/kg and 0.25mg/kg. The control group were given water without ginsenoside. Every day, the amount of food was recorded, and the food and mice were weighed every week. The utilization rate and the weight of ginsenoside-Rb$_1$, Rg$_1$ were also recorded. As shown in table 1, the amount of ginsenoside Rb_1 and Rg_1 was 28.6, 56.1, 27.4, 53.9 mg/kg per day. After giving the mice ginsenoside-Rb_1, Rg_1, their weight increased rapidly. Brain weight and pallium thickness also increased rapidly. In the passive evade experiment, Rb_1 28.6mg/kg group and 53.9mg/kg group decrease more than other groups (Zhang Juntian et al., 1988; Yangying et.al, 1994) (Table 1).

Table 5: Effect of ginsenoside-Rb_1, Rg_1 on mice's brain nerve grow

Group	Ginsenoside	Brain weight	Pallium thickness
Contrast	-	5.6±0.4	1.8±0.1
Rb_1	28.6	5.9±0.4	1.9±0.1
Rb_1	56.1	6.4±0.7	2.0±0.2
Rg_1	27.4	6.5±0.4	2.1±0.1
Rg_1	53.9	6.5±0.5	2.0±0.2

4.1.2 EFFECT OF GINSENOSIDE ON ISCHEMIA MODEL ANIMAL

Tests were done to determine whether ginseng has an effect on nutritional activity and protecting activity on the brain? The results showed that ginseng had some effect on increasing nervous growth factor, and when some nervous cells were aggressively being destroyed, ginsenoside was able to restore it. Thus, ginseng had some effect on protecting ischemia of heart.

An animal model was made using congreal dog's neck artery. In the first week, the animals were given red ginseng powder the animal passive study obstacle was adjusted, each dose 0.6-1.5g/kg per day. The action latent period was prolonged, and nerve cells in seahorse CA_1 area increased more than the contrast group, showing that nerve cells were protected. Orally taking coarse ginseng saponin (CGS) (50-100mg/Kg) had the same protection.

One week before being making the ischemia model, inject ginsenoside-Rb_1 once a day. By comparing and contrasting the results with the control group, it was clear that there was protection of seahorse cell, but no Rg_1 or R_o (Figure 6).

By giving the animals low-density ginsenoside-Rb_1, damage by free OH was prevented, showing that Rb_1 had provided protection. When brain ischemia threatened a human's life, even injecting or orally taking Rb_1 was useless because Rb_1 had not reach brain. When ischemia 3.5 min was taken Rb_1 60-600mg/kg per day, the group could retain some seahorse cell but none were alive in contrast group.

A permanent block animal model by SH-SP mouse was made. After being blocked two hours, inject Rb_1 0.006-6mg/kg four weeks. It can adjust mouse study ability. Compared to the control group, the blocked two hours group shows improved study ability in four weeks, brain block decrease, and brain area nervous optics neuron decrease. This brain block model shows ginsenoside-Rb_1 has protection action.

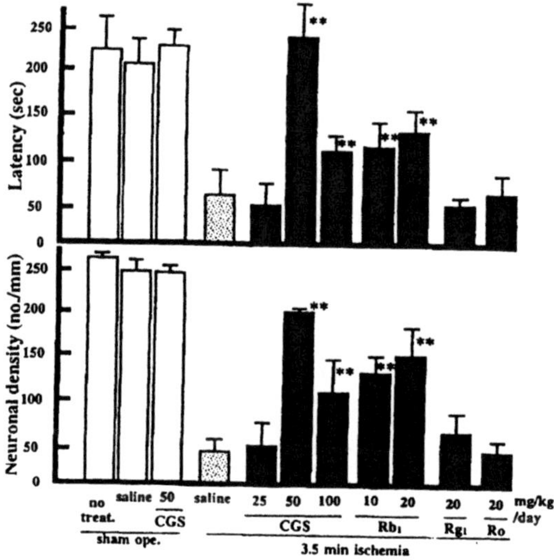

Figure 6: Effect of ginsenoside on ischemia animal model.

4.1.3 EFFECT ON PLATELET AGGLUTINATION

ADP, Collagen, epinephrine, arachidonic acid, and A2(TXA2) are provocation of platelet agglutination. Low-density ginsenoside-Rg_1 had a refrained effect on them.

After giving 50mg ginsenoside-Rg_1 to six healthy men, blood was drawn after three hours. ADP, collagen, epinephrine, arachidonic acid, A2 (TXA2) were measured. The results showed platelet agglutination had been restrained (Table 6). An experiment with Rg_3 had the same effect (Hikokichi).

Table 6: Effect of ginsenoside-Rg_1 given one time on platelet aggregation

Coagulation	Capabilities of platelet aggregation			
	Before administration		3h later	
Collagen	56	15	37	20
Epinephrine	65	11	19	30
STA₂	70	6	43	21

In red ginseng, the ginsenoside-Rg_1 is comparatively low. Recently, it was found that in gastric juice, 0.1N HCl, the C_{-20} sugar chain was cut off from ginsenoside-Rb_1, Rb_2, Rc, and Rd when taken orally. Though the effects on platelet aggregation are not significant of ginsenoside-Rb_1, Rb_2, Rc and Rd, which are higher in red ginseng, it is still noteworthy that the ginsenoside-Rg_3 could restrain platelet aggregation as Rg_1 does, which generate from them (Figure 7).

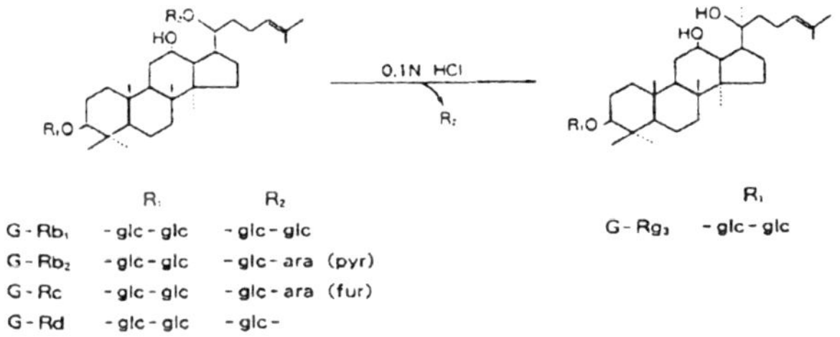

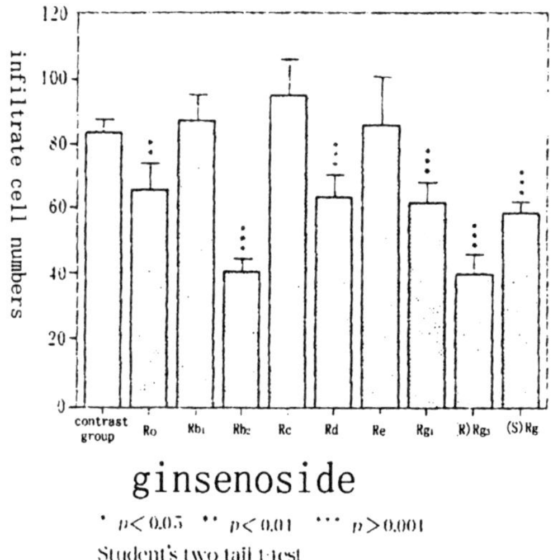

Figure 7: Reactions showing restrained platelet aggregation.

4.2 ANTI-CANCER EFFECT OF GINSENOSIDE-RG$_3$

4.2.1 DIFFERENT GINSENOSIDES HAMPER MELANIN-B16-BL6 CELL INFILTRATE

Before cancer cells are transferred to the other organs, it infiltrate around. Melanin-B16-BL6 cell had been processed by various ginsenosides and then added to liver organs. The Transwell cell cultivate method was used in melanin cells. After five hours, the experiment showed restrain infiltrate cell activity by ginsenoside-Rb$_2$, 20-ginsenoside-Rg$_3$, and 20-ginsenoside Rg$_3$, but not Rd, Rg$_1$, Ro (Yangying et al., 1994) (Figure 8).

Figure 8: Various ginsenosides hamper melanin-B16-BL6 cell and infiltrate around organ cells.

4.2.2 ANTI-CANCER ACTIVITY OF GINSENOSIDE-RG$_3$

An inoculation was attached to the rats' belly after nine days, containing ip ginsenoside-Rg$_3$. The cancer cells were measured at different times (each one 2×10^4 cells). Each period of time phase cell rates were recorded (Tables 7, 8, 9).

Table 7: Effect of various dosages of ginsenoside-Rg$_3$ on S180 cells

Time phase	Blank	3mg/kg	6mg/kg
G1/10	84.53	3.78	18.12
S	6.84	46.67	35.92
G2/M	8.62	49.56	45.95

Note: The data in Table 7 is the result giving ginsenoside-Rg$_3$ for 24 hours.

Table 8: Effect of ginsenoside-Rg$_3$ on different tumor cells increasing period.

Time phase	S180		Heps		EAC	
	Blank	3mg/kg	Blank	3mg/kg	Blank	3mg/kg
G1/10	84.53	3.87	93.36	83.55	78.32	22.72
S	6.84	46.67	4.98	13.80	20.23	26.26
G2/m	8.62	45.96	1.66	2.65	1.45	51.02

Note: The data in Table 8 is the result giving ginsenoside-Rg$_3$ for two hours.

Table 9: Effect of ginsenoside-Rg$_3$ on different time of EAC cells increasing period

Time phase	Blank	2 hr.	4hr.	8 hr.	24 hr.
G1/10	78.32	26.75	31.78	28.74	16.90
S	20.23	16.82	14.48	7.31	33.20
G2/M	1.45	56.44	53.74	63.95	49.90

Note: The data in Table 9 is ip ginsenoside-Rg$_3$ 3mg/kg.

According to the Tables 7-9, ginsenoside-Rg$_3$ had the following effects during the cancer cells increase period:

(1)   Main active time phase was G$_2$. The cells restrain rate in this period increased. It showed that ginsenoside–Rg$_3$ restrained cancer cell pre-divide, raw material tubulin, and ATP synthesis, so it decreased the speed of cancer cell growth.

(2)   The effect on different tumor cell increase during this period was on S$_{180}$G$_1$,S, and G$_2$. In G$_1$ period, cell increase rate was lowered. It showed that ginsenoside-Rg$_3$ restrained synthesis of RNA and protein. In S period, cell rate increased. It showed that Rg$_3$ restrained DNA synthesis. EAC active in the G$_2$ period nearly restrained Heps cell increase.

(3)   As S$_{180}$, ip ginsenoside-Rg$_3$3mg/kg had more restrain activity.

(4)   As EAC, giving medicine for two to eight hours, after 24 hours, showed restrain activity, especially increase in G$_1$ and S period. There was a decrease in G$_2$ period.

4.2.3 GINSENOSIDE-RG$_3$ ANTI-TRANSFER ACTIVITY

When an inoculation with cancer cells was in armpit, ginsenoside-Rg$_3$ was given orally. With the high dose group, it showed 59.41% and 67.82% restrain rate in B16 melanin tumor transfer and lung cancer cell transfer (Tables 10 and 11).

Table 10: The effect of oral-taking ginsenoside-Rg$_3$ on B16 melanin tumor transfer.

Group	Dose	Scheme	Animal number	Animal weight	Lung Clones	Restrain rate(%)
Ginseno-	3.00	Po×10qod	10/10	18.2/21.4	20.5±10.48	59.41[**]
side	0.60	Po×10qod	10/10	18.7/22.3	22.6±10.37	55.25[*]
Rg$_3$	0.12	Po×10qod	10/10	18.4/21.5	37.2±13.4	26.34[*]
CGS	100.00	Po×10qod	10/10	18.3/21.9	35.7±12.7	29.31[*]
CTX	30.00	Ip×7qd	10/10	18.6/20.2	7.30±7.07	85.54[**]
Contrast		Po×10qod	20/20	18.4/22.8	50.5±18.9	

* P<0.05, ** P<0.01 data in tab is average in three experiments

Table 11: The effect of oral-taking ginsenoside-Rg$_3$ on lung cancer cell transfer.

Group	Dose	Scheme	Animal number	Animal weight	Lung Clones	Restrain rate(%)
Ginseno	3.00	Po×10qod	10/10	18.5/21.4	2.13±2.61	67.82[**]
side	0.60	Po×10qod	10/10	18.4/21.6	2.73±3.23	58.76[**]
Rg$_3$	0.12	Po×10qod	10/10	18.7/21.2	4.19±3.21	26.34[**]
CGS	100.00	Po×10qod	10/10	18.6/21.3	4.57±4.50	29.31
CTX	30.00	Ip×7qd	10/10	18.7/20.4	0.50±1.00	85.54[**]
Contrast		Po×10qod	20/20	18.6/22.4	6.62±4.92	

** P<0.01 data in tab is average in three experiments

4.2.4 EFFECT OF GINSENOSIDE-RG$_3$ ON TUMOR CELL DIE

In study of cell increase dynamics, "die summit" with Ginsenoside-Rg$_3$ was observed. The rate of cell death was determined (Fuli et al., 1997).
- (1) Taking 3mg/kg/day had significant activity on S$_{180}$. Its rate was 3.16%, 1.8 times to 6mg/kg.
- (2) Taking 3mg/kg two, four, eight and 24 hours, the rate of cell death was 3.25%, 5.13%, 4.26%, and 10.08%, respectively.
- (3) Taking orally 3mg/kg two hours, the rate of cell death of S$_{180}$, Heps, and EAC was 19.64%, 5.26%, 2.96%, respectively.

4.2.5 PROMOTE IMMUNIZATION ACTIVITY BY GINSENOSIDE-RG$_3$

Ginsenoside-Rg$_3$ promoted rat's immunization. Taking orally 0.6mg/kg improved Lewis rat NK cell and IL-2 active is 46.41% and 3122±735 (Tables

12 and 13).

Table 12: The effect of ginsenoside-Rg$_3$ on Lewis lung-cancer mice NK cells activity.

Group	Dose	Scheme	OD (X)	NK activity (%)
Ginsenoside	3.00	Po×10qod	0.378	42.29[**]
Rg$_3$	0.60	Po×10qod	0.351	46.41[**]
	0.12	Po×10qod	0.368	43.82[**]
CGS	100.00	Po×10qod	0.381	41.83[**]
Contrast		Po×10qod	0.525	19.84

[**] P<0.001,OD =0.655

Table 13: The effect of ginsenoside-Rg$_3$ on Lewis lung-cancer mice IL-2 activity

Group	Dose	Scheme	Animal numbers	IL-2 activity/cpm (X±SD)
Ginsenoside	3.00	Po×10qod	6	2836±326[**]
Rg$_3$	0.60	Po×10qod	6	3122±735[**]
	0.12	Po×10qod	6	2414±539[**]
CGS	100.0	Po×10qod	6	2785±255[**]
Contrast		Po×10qod	10	1351±127

[**]P<0.01

4.3 PROTOPANAXADIOL-TYPE GINSENOSIDES METABOLISM IN STOMACH AND IN BOWEL AND ITS PRODUCT

4.3.1 TRANSFORMATION OF GINSENOSIDE IN STOMACH

Take 20s-panaxdiol orally, metabolism in stomach (0.1N HCl), C-20 glucose had been. Ginseng, especial red ginseng, had a sufficient amount of ginsenoside-Rb$_1$, Rb$_2$, Rc, and Rd. These ginsenosides have anti-cancer activity themselves but indirectly. However, by stomach liquid, it turned into Rg$_3$. Rg$_3$ had the same effect on platelet agglutination just like Rg$_1$. Rb$_1$,Rb$_2$,Rc, and Rd are the pre-body of Rg$_3$ which is an interesting active compound.

4.3.2 GINSENOSIDE-RB$_1$ METABOLISM IN BOWEL

When ginsenoside-Rb$_1$ in human excrement and urine is mixed with bacteria, the transformation occurs. First, it turned into ginsenoside-Rd and then into Rd and F$_2$. Finally, it turned into Compound K. After cultivate 24 hours, the glucose in C-20 had been cut turn into agrocon 20s-protopanaxadiol. Filtering it with bacteria, confirmed it as *Eubacterium*, SP-44. This bacteria had same effect on Chinese thorowax saponin.

109

After taking ginsenoside $-Rb_1$, it hadn't been examined out Rb_1 in blood, but examined compound K. After taking compound K, its density increase with time, it can maintain over eight hours. Rb1 had effect on lung cancer cells due to Compound K's effect. So Compound K is an anti-cancer agent (Koyoichi) (Figure 9).

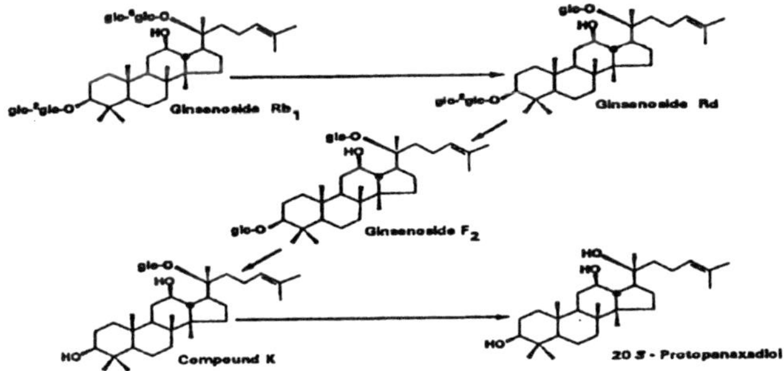

Figure 9: Reaction showing Compound K, an anti-cancer agent.

5. CONCLUSION

The work presented in this report is still in the early stage of research on the metabolism of ginsenoside. It demonstrates the significance of separation of ginsenoside from metabolism product. To determine whether each ginsenoside has an individual difference requires further studies. Metabolism product and bioactivity will be an important topic in the 21st century.

REFERENCES

Fuli et al. Studies on anti-tumor mechanism of ginsenoside-Rg₃, the second seminar of natural medicine resource, 1997, 5, 114.

Kobashi Koyoichi, Ginsenoside metabolism in bowel and absorption. The Ginseng Review, No. 26:P90-91.

Li xianggao Studies on extraction and bioactivity of new components from ginseng, P135-P138, Jilin Agricultural University.

Oura Hikokichi, Physiology and pharmacology activity of ginsenoside, The Ginseng Review, No. 26:P35, P50.

Yangying et al., Discuss on Rb_1 and Rg_1 promote wise activity mechanism, ACTA PHARMACEUTICA SINICA, 1994, 29:P240.

Zhang juntian et al., Effect of Ginsenoide-Rg_1 and Rb_1 on center nerve system transmitter receiver and brain synthesis protein 1988, 32:P12.

Chapter 12

ANTIBACTERIAL SYNERGY IN RUBRICINE: AN EXTRACT FROM THE ROOTS OF *ARNEBIA EUCHROMA* A CHINESE MEDICINAL HERB

SPENCER A BENSON, BRIAN P. HIGGINS, CHI S. CHAE, AND YUAN LIN
University of Maryland, College Park, MD 200742

Abstract: The use of natural products for medicinal purposes can be traced back more than 3000 years in Traditional Chinese Medicine (TCM). TCM herbalists generally use a combination of compounds (herbs) to treat the patient and restore balance. One agent used by herbalists is an organic root extract from *Arnebia euchroma* (Zi Cao). It is traditionally used as a healing agent for wounds, burns, abrasions, and other dermatitis related problems. Extracts from *Arnebia euchroma* and related *Boraginaceae* plant species contain a variety of therapeutic activities, which include an antimicrobial activity that is active against a variety of bacterial and fungal species. To characterize the antimicrobial activity (ies) present in root extracts from *Arnebia euchroma*, we used *Staphylococcus aureus* and *Bacillus subtilis* as a model system. When the *Arnebia* roots are crushed and extracted with an organic solvent, a concentrated red extract, rubricine, is obtained. Rubricine contains the naphthoquinine compound shikonin and five structurally related compounds. Each compound differs from shikonin by a single R-group moiety. Using our model system, we showed that rubricine has both bacteriostatic and bactericidal activities, each of the compounds present in rubricine is active, rubricine is effective against multiple-drug resistant (MDR) clinical isolates, and shikonin acts synergistically with each of the other compounds to enhance their antibacterial activity. One explanation for the multiplicity of related compounds in rubricine is that plants naturally use TCM type strategies that capitalize upon mixtures of active agents to address complex microbial challenges. These findings suggest that searches for new compounds effective against emerging MDR organisms might include assays for combinatorial activities that build about historical TCM knowledge.

Yuan Lin (ed.), Drug Discovery and Traditional Chinese Medicine: Science, Regulatory and Globalization, 111-123.
©2001 *Kluwer Academic Publishers. Printed in the Netherlands*

1. INTRODUCTION

The emergence and spread of resistant strains of bacteria and fungi increasingly compromise the use of existing anti-microbial agents for the treatment of infections and diseases. The worldwide increase in bacterial infections that fail to respond to conventional antibiotic therapy due to multiple drug resistance (MDR) is of increasing concern for all (Levy, 1992, ASM, 2000). Of special concern are aggressive infections that do not respond to conventional antibiotics due to MDR phenotypes. As such, there is an important need to identify new approaches for finding and characterizing new antimicrobial substances. One way to meet this need is to capitalize upon the historical knowledge of herbal products used to treat infections.

Natural (herbal) products have been used as medicinal agents in Traditional Chinese Medicine (TCM) for many centuries. A fundamental tenant of TCM is that herbal combinations yield benefits that exceed the additive effects of individual active components (Lin and Yuan, 1997). Herbalists formulate mixtures of agents to restore and maintain balance among the body systems (Yuan and Lin, 2000). The mixture usually combines an agent that provides the desired therapeutic activity with agents that accentuate the potency of the first agent, treat accompanying symptoms, moderate harshness or toxicity, guide the medicine to the proper target, or exert a harmonizing effect. This holistic approach is in contrast to western medical approaches that generally use single agents for treatment of acute conditions. The "Magic Bullet" approach is the legacy concept pioneered by Paul Erlich (Erlich, 1908) that helped to establish the modern pharmaceutical industry.

Among the many uses of herbal products is their application for preventing and controlling infections (Tang, 1992). Organic extracts from the roots of members of the *Boraginaceae* family (e.g. *Alkanna tinctora, Lithospermum erythrorhizon, Arnebia euchroma*) have historically been used in Europe and China as medicinal agents for wound healing, in the treatment of burns, and as a dye for textiles. In modern times, ointments containing extracts from these plants have become popular treatments for other skin conditions, such as allergic dermatitis. In addition, burn, wound, and dermatitis healing extracts from this family of plants have biopharmaceutical properties that include antibacterial, anti-fungal, anti-inflammatory, and anti-thrombic activities (reviewed in Papageorgiou et al., 1999). Typically, these plant (root) extracts contain the antiomeric naphthoquinones, shikonin or alkannin plus several structurally related chemical agents. Organic solvent extraction of the roots of *A. euchroma* (a.k.a. purple root, reddish-purple root, or zi cao) yields a deep red extract, termed rubricine. Rubricine contains shikonin (S), and five related naphthoquinone compounds, acetyl-shikonin (AS), teracryl-shikonin (TS), β-hydroxyisovaleryl-shikonin (HIVS), isobutyl-shikonin (IBS), and β,β-dimethylacryl-shikonin (DMAS) (Figure 1). The individual components can be separated by preparative HPLC or thin layer chromatography (Brigham, et al., 1999) or chemically synthesized (Papageogiou, et al., 1999).

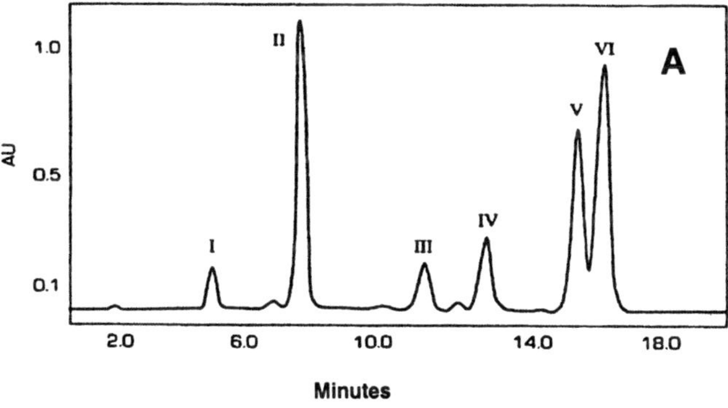

Figure. 1. HPLC chromatogram of rubricine extracted from *A. euchroma*. *Arnebia euchroma* roots were pulverized and extracted with methanol. After drying, the extract was dissolved in chloroform and individual agents were separated by HPLC (Brigham et al. 1999). The various peaks were identified by A_{520} adsorption. Structures were assigned based on their mass-spectrometer profiles. Peaks (I to VI) in panel A correspond to the structures in panel B.

2. RUBRICINE IS ACTIVE AGAINST A WIDE VARIETY OF MICROBES

Using a simple disk assay for anti-microbial activity we, showed that rubricine inhibits the growth of a variety of bacterial and fungal species (Table 1). Rubricine is active against multiple-drug resistant (MDR) *Staphlococcus aureus* and *Enterococcus faecali* strains, as well as clinical isolates, which include vancomycin resistant *Enterococcus faecali* (VREs), quinolone resistant (QRS) and methicillin resistant (MRSA) *Staph. aureus* strains (Table 1). Rubricine is more active against Gram-positive bacteria compared to Gram-negative bacteria. This difference in activity is due to the permeability barrier of the outer membrane. Mutations that alter outer membrane permeability, such as the *imp* (Sampson et. al., 1988) or *rfa* mutations (Nakaido, 1996), confer sensitivity to rubricine (Table 1). Rubricine is also active against yeast and fungi species, including animal and plant pathogen species (Table 1). Other non-enteric Gram-negative bacterial and fungal plant pathogens are reported to be sensitive to extracts from *Arbenia euchroma* and individual compounds (Brigham et. al., 1999).

Table 1. Activity of Rubricine Against Bacteria and Fungi

Organism	Organism type	Growth Inhibition
Bacillus subtilis 168	Gram +	Yes
*Enterococcus fae*calis (wild type) ATCC49149	Gram +	Yes
*Enterococcus fae*calis (MDR) ATCC 515175	Gram +	Yes
*Enterococcus fae*calis clinical isolates VRE[1]	Gram +	Yes
Staphylococcus aureus (wild type) ATCC 29213	Gram +	Yes
Staphylococcus aureus (MDR) ATCC 27659	Gram +	Yes
Staphylococcus aureus clinical isolates[2]	Gram +	Yes
Staphylococcus aureus grlA, gyrB	Gram +	Yes
Streptococcus pyrogenes	Gram +	Yes
Streptococcus epidermis	Gram +	Yes
Streptococcus pneumoniae ATCC 6301	Gram +	Yes
E. coli K12	Gram -	No
E. coli K12 *imp* (a hyperpermeable strain)	Gram -	Yes
Proteus vulgaris	Gram -	Slight
Pseudomonas aeruginosa	Gram -	No
Salmonella typhimurium LT2	Gram -	No
Salmonella typhimurium rfa (a hyperpermeable strain)	Gram -	Yes
Trychophyton ruben	Fungi	Yes
Fusarium proliferatum	Fungi	Yes
Pistoris pastis	Yeast	Yes
Saccharomyces cerevisiae	Yeast	Yes

Activity was tested with rubricine (20-40µg) using a standard disk sensitivity test.
[1]Five clinical isolates were tested. [2]More than 20 independent antibiotic resistant isolates were tested from both the US and China.

3. RUBRICINE IS BOTH BACTERICIDAL AND BACTERIOSTATIC

When a bacterial culture is treated with rubricine, we observe two responses, a rapid killing effect that reduces the number of colony forming units (CFU) by several orders of magnitude and cessation of growth as measured by optical density (Figure 2). A small percentage of the cells in the culture remain viable in the presence of rubricine and recover the ability to grow after several hours (Figures 2, 3). These cells exhibit full sensitivity to rubricine when subsequently tested using disk sensitivity or minimum inhibitory concentration (MIC) assays in liquid media. When the amount of rubricine is increased, the delay before the population resumes growth is concomitantly increased (Figure 3). This suggests that the release of the growth is due to breakdown or metabolism of rubricine. Treatment with rubricine does not result in cell lysis as evidenced by the lack of a reduction in OD following the addition of rubricine (Figures 2, 3). Furthermore, light microscopy of the moribund culture reveals that cells are intact and that there is an absence of noticeable cell detritus.

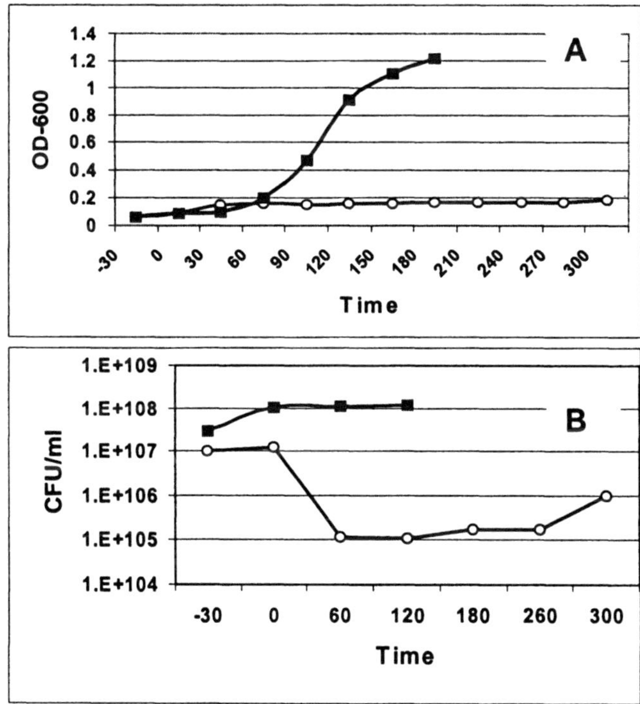

Figure 2. The antibacterial activity of rubricine. A growing culture of *Staphylococcus aureus* ATCC 27659 was treated with rubricine (2μg/ml) in LB media and OD_{600} was monitored (panel A) as well as colony forming units (CFU) on LB media (panel B). Closed squares ■ are without rubricine and the open circles ○ are with rubricine. Rubricine was added at t=0.

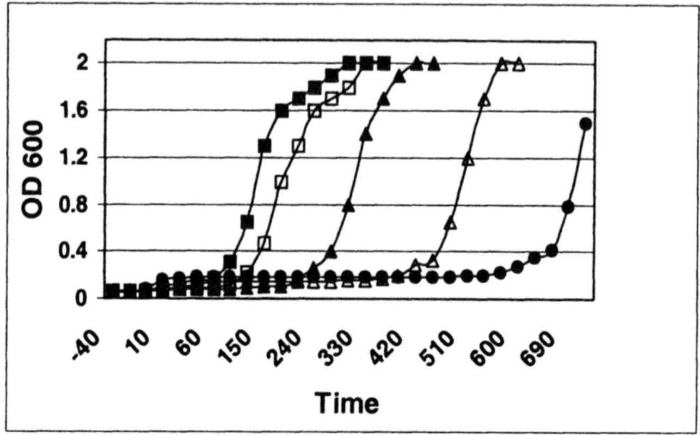

Figure 3. Release time versus amount of rubricine. A growing culture of *Staphylococcus aureus* ATCC 27659 was treated with rubricine in LB media and OD_{600} was monitored. ■ = 0 µg/ml rubricine, ◻ = 0.5 µg/ml rubricine, Δ = 1 µg/ml rubricine, ▲ = 2µg/ml, ●= 3µg/ml. Rubricine was added at t=0.

4. ALL AGENTS PRESENT IN RUBRICINE ARE BIOLOGICALLY ACTIVE

To determine if individual compounds present in the extract have antibacterial activity, we used a standard disk assay (Figure 4) and determined MICs in Luria broth (Silhavy et. al., 1984) (Table 2). All of the agents inhibit bacterial growth in a dosage dependent manner (Figure 4). For rubricine and each of the individual agents, the size of the zone of growth inhibition increases with increasing amounts, until a plateau is reached where additional amounts of compound result in no further increase in zone size (Figure 4). For rubricine and the individual agents, concentrations of 20-40 µg per disk give a maximum zone size. Rubricine is more active than the majority of the individual components (Table 2, Figure 4).

The concentration of rubricine and the individual components necessary to inhibit bacterial growth (MIC) in liquid media (LB) are in the 0.5-6.0µg/ml range (Table 2). These MIC values are similar to those of many commonly used antibiotics, such as the tetracyclines. The MIC for rubricine (0.5 µg/ml) is lower than expected given the MICs of the individual components (0.5 to 6.0 µg/ml). There are differences in the apparent relative antibacterial activities in the liquid MIC and disk diffusion tests for several of the agents (e.g. compare the results in Table 2, with those in Figure 4). For example, DMAS has a low MIC (0-2.0 µ/ml) and produces a small zone of inhibition. Conversely, shikonin has one of the highest MIC (4-8 µg/ml) and produces a large zone of inhibition. Presumably, the differences seen with individual well as the nature of the assays themselves. Different bacterial species give similar, but not identical, zone of inhibition sizes and MIC values. Gram negative species tend to have slightly higher MIC ranges, in the 4-8 µg/ml.

Rubricine has a low MIC (0.5-1.0) and gives a relatively large zone of inhibition (Table 2, Figure 4). The two most active agents (e.g. those with the lowest MIC) acetylshikonin (AS) and ββ-dimethylacrylshikonin (DMAS), compromise less than 60 percent w/w of the rubricine mixture (Table 2), yet rubricine has an MIC comparable to each of these single agents and produces large zones of inhibition. The low MIC and large zone of inhibition size for rubricine suggests that individual components might interact in a synergistic manner.

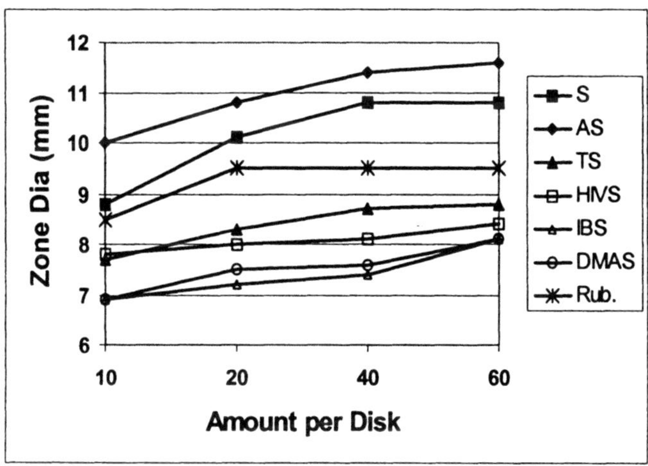

Figure 4. Antibacterial activity for rubricine and individual agents. Paper disk (6mm) containing various amounts of the agent were placed on a lawn of *Staphylococcus aureus* ATCC 27659. After 12-14 hr incubation the size of the zone of growth inhibition was measured in mm. The agents are rubricine (Rub.), shikonin (S), acetyl-shikonin (AS), treachery-shikonin (TS), β-hydroxyisovaleryl-shikonin (HIVS), isobutyl-shikonin (IBS), and β,β-dimethylacryl-shikonin (DMAS).

Table 2. Characteristics of Rubricine from *Arnebia euchroma.*

Agent	Percent Composition in Extract	MIC (µg/ml)	Zone of Growth Inhibition (mm)
Rubricine		0.5 - 1.0	9.5
Peak I (Shikonin)	3	4.0 - 5.0	10.1
Peak II (AS)	24	0.5 - 1.0	10.8
Peak III (TS)	5	4.0 - 6.0	8.8
Peak IV (HIVS)	9	4.0 - 6.0	8.0
Peak V (IVS)	26	2.0 - 4.0	7.2
Peak VI (DMAS)	35	0.0 - 2.0	7.5

MIC and zone of inhibition data is for *Staphylococcus aureus* ATCC 27659. A MDR strain. Similar results are obtained with other Gram-positive organisms.

To test if there is synergy between individual agents, we used a disk test to determine whether the parent compound shikonin affected the activity of the other structurally related compounds. The basic procedure involved adding shikonin (40µg) to a series of standard blank paper disks and then adding 0-60µg of the second agent X (X=AS through DMAS) to the same disks. A control set of disks containing only X (0-60µg/disk) was prepared, in addition to a control disk that contained only shikonin (40µg). All disks were placed on a lawn indicator of bacteria (~ 10^7 cells), suspended in minimal top agar. The plates were incubated overnight at 37° and scored 12-14 hr later. If the two agents, shikonin plus X, operate in an identical manner then the addition of shikonin should have no effect since shikonin is present above a maximum effective concentration. At concentrations greater than 20 µg/disk, both components are at concentrations within the plateau range of the assay (Figure 4). If shikonin and the second agent (X) operate independently, then the size of the zone of inhibition dose response line for the disk containing shikonin plus X should parallel the dose response line for compound X. Again, since shikonin is added in an amount above its maximal effective concentration its contribute would remain constant. If compounds shikonin and X act in a synergistic manner then the zone of inhibition dose response for Shikonin A plus X should rise. When a constant amount of shikonin was combined with increasing amounts of any of the other compounds (AS through DMAS), the zone size increased in a near linear manner. Representative data is shown in Figure 5 for shikonin plus DMAS. Similar results were obtained for each of the other agents present in the rubricine and rubricine itself. In each case, the addition of shikonin significantly increased the antibacterial activity of the second agent (Figure 6).

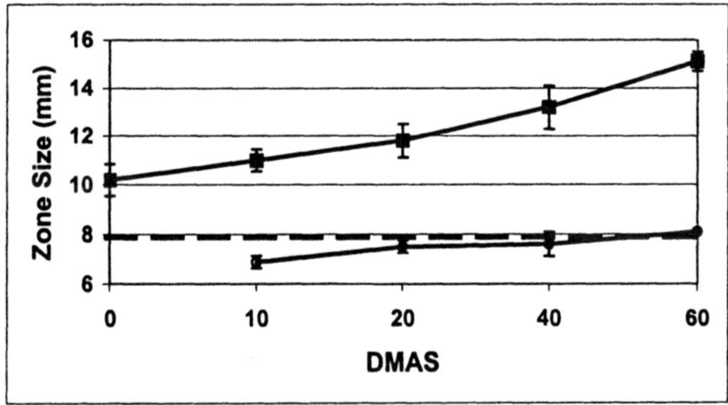

Figure 5. Synergistic interaction of shikonin and DMAS. Activity was measured as described in the legend to Figure 4. Shikonin (60µg) was added to paper disk containing the indicated amount of DMAS (x-axis). The dashed line shows the expected effect if the activities are additive. Abbreviations are as described in the legend to Figure 4. Error bars represent standard deviation, N= 11.

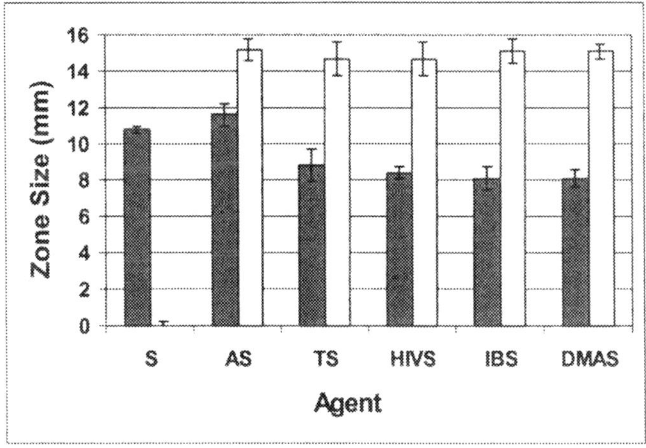

Figure 6. Synergy of shikonin with other agents present in rubricine. Activity was measured as described in the legend to Figure 4. Shikonin (60μg) was added to paper disk containing individual an agent (40μg). The solid bars show the zone size in without shikonin and the open bars show the result with shikonin. The errors show the standard deviation of 5-8 replicate experiments. Abbreviations are as described in the legend to Figure 4. Error bars represent standard deviation, N= >10.

5. DISCUSSION

Rubricine contains a series of structurally related naphthoquinone compounds that exhibit broad-spectrum antimicrobial activity, both individually and in combinations. Rubricine is unique in that it has both antibacterial and anti-fugal activities. This suggests that the cellular target(s) on which rubricine acts are conserved across the procaryotic-eucaryotic evolutionary domains. Alternatively, the agents may have low target specificity and kill by a general mechanism rather than inactivating a specific target or cellular pathway. A mechanism of general disruption of metabolism seems unlikely, given that rubricine exhibits both bactericidal and bacteriostatic activities. The finding that a small percentage of cells with the population are able to resume growth after many hours of exposure is inconsistent with a general disruption mechanism. Since rubricine has both bactericidal and bacteriostatic activity, a more plausible explanation is that rubricine interacts with more than one cellular target or that the target is involved in more then one cellular pathway (e.g. replication and transcription). Still, it is not possible at this time to distinguish between these possibilities, nor fully eliminate a general disruption mechanism.

The reason why a small sub-population of cells within the growing culture are arrested in growth but not killed is unknown. The simple explanation that they represent resistant mutants is not viable since these cells exhibit normal sensitivity to rubricine and the individual agents when subsequently tested. Moreover, true resistant mutants would grow in the presence of rubricine, while this small population of cells retains sensitivity to rubricine's ability to stop cell growth. One explanation for how this sub-fraction of cells within the population can be transiently resistant to the bactericidal effect of rubricine but not the bacteriostatic effect is that they are in a certain physiological state at the time of exposure that results in a stoppage of cell growth but not death. Since the bulk of the cells are presumably not in this state, growth is not blocked and the result is cell death. We observed that, following an addition of rubricine, there is a slight increase in OD. However, the number of CFU decreases during this same time period. Together, these results suggest that cell growth may be required for killing. Further tests will be necessary to confirm this hypothesis.

The time before these transiently resistant cells resume growth is a function of dosage (Figure 3). Two explanations for the delay are i) the number of transiently resistant cells is inversely related to the concentration of rubricine added or ii) the active agents are broken down. Since we do not see increased levels of cell killing with increasing concentrations, the second explanation is favored. At this time, we cannot distinguish between rubricine being metabolized versus chemical breakdown. The compounds that make up rubricine have long half-lives under a variety of chemical and physical conditions (Cho et al., 1999).

The two biological assay systems we used have different characteristics although both monitor stoppage of cell growth or death. Given the hydrophobic nature of the compounds and the size and structure of the R-groups. it is expected that the various compounds would have different diffusion and solubility characteristics and thus yield different zone sizes. The characteristically small zones of growth inhibition for the larger hydrophobic compounds may reflect physical properties of the molecule rather than biological activity. This appears to be the case for DMAS, which gives a small zone of inhibition (Table 2) but has a low MIC. The concentrations used in the MIC assay are within the solubility limits of the compound in aqueous solutions, whereas the amount applied to the disk is many folds higher. The importance of using multiple types of assays in the characterizing herbal activities is underscored by these findings. By combining both MICs in liquid media and disk sensitivity tests, we were able to obtain truer profiles of biological activity.

The antibacterial activity of all of the individual compounds (AS through DMAS) is enhanced by the presence of the parent compound shikonin. The degree of enhancement is significant and most prevalent for the larger compounds. The mechanism for the synergy is not known. One possibility is that shikonin increases the uptake of the larger agents. We favor the alternative explanation that the two agents act on different target molecules or at different steps within a common pathway and results in an amplification of the antibacterial activity. Synergy between compounds may account for high potency of rubricine. However, the amount of shikonin present in rubricine is low (Table 2), which suggests that other beneficial interaction may occur within the mixture.

Given the planer structure of the compounds and reports that shikonin affects the activity of topoisomerases (Fujii et al., 1992; Plyta et al., 1998) a feasible target pathway for the compounds is replication and/or transcription. This pathway could account for the ability of the compounds to arrest cell division and also cause in cell death. Furthermore, the effect on transcriptions is consistent with the wide range of biological effects (Papageorgious et al. 1999) in both procaryotic and eucaryotic cells. Further experiments will be necessary to test this hypothesis. Mutations in *Staphylococcus aureus* (*gyrA* and *gyrB*) that block gyrase functions (Fournier and Hooper, 1998; Kaatz and Seo, 1998) do not alter sensitivity to rubricine, suggesting that gyrase is unlikely the target molecule. If so, the compounds interact in a manner different than the quinolone antibiotics. Our attempts to directly isolate mutants that are resistant to rubricine, shikonin, or acetyl shikonin have been unsuccessful, which supports the hypothesis that different agents present in rubricine act on different target molecules. The inability to find mutants resistant to individual agents suggests that they may also interact with more then one target. Further work is necessary to test if this is indeed the case.

6. CONCLUSION

During the last decade, the search for new antimicrobial drugs has been revolutionized by high through put screening, combinatorial chemistry, structural analysis of target molecules, genetic screens for essential genes, and increased screening of natural products. In the latter case, the goal has been to identify the active agent and then focus further characterizations and development of single agents. While this strategy has seen much success, it also results in a situation where resistant mutants can quickly arise and compromise the usefulness of the agent. An additional strategy for drug discovery and development is to build upon the wealth of empirical information on herbal formulations. Many TCM formulations have wound healing properties which, may be associated with the antibacterial activities and provide a pre-screen for potential new antimicrobials. Rubricine is an example of how this approach is valuable. Characterization of the activities of the herbal mixture and the individual components suggest they represent novel antibacterial agents. Moreover, the finding that shikonin enhances that activity of the other agents indicates that screens that focus on single agents may overlook potentially important new drugs and drug combinations that are more effective and reduce the occurrence of resistant organisms. Lastly, the finding that *A. euchroma* and presumably other members of the *Boraginaceae* family use a natural combinatorial strategy in producing antimicrobial agents raises the possibility that the strategies used by herbalist in combining agents to holistically treat patients may be grounded in a basic strategy of nature. The reason why plants of the *Boraginaceae* family produce an abundance of napthloquinone compounds in root tissue is not known. However, it seems likely that they produce these secondary metabolites as a defense to microbial infections since expression of these metabolites is induced by exposure to fungal compounds and bacterial pathogens (Brigham et al. 1999).

REFERENCES

American Academy of Microbiology. Antimicrobial resistance: an ecological perspective. Washington D.C.: ASM-Press, 2000

Brigham L., Michaels P.J., Flores H.E. Cell specific production and antimicrobial activity of napthoquinones in roots of *Lithospermum erythrorhizon*. Plant Physiology 1999; 119:417-428.

Erlich P. Modern Chemotherapy Beitrage zur experimentellen Pathologie and Chemotherapie, 1908 In *Milestones in Microbiology*, Brock, T. B. Washington DC: ASM-Press, 1999: 167-202

Fournier B. and D.C. Hooper. Mutations in topoisomerase IV and DNA gyrase of *Staphylococcus aureus*: novel pleiotropic effects on quinolone and coumarin activity. Antimicrobial Agents and Chemotherapy 1998; 42:121-128.

Fujii N., Yamashita Y., Arima Y., Nagashima M., and Nakano H. Induction of topoisomerase II-mediated DNA cleavage by the plant naphthoquinones plumbagin and shikonin. Antimicrobial Agents and Chemotherapy. 1992; 36:2589-2594.

Kaatz G.W. and S.M. Seo. Topoisomerase mutations in fluoroquinolone-resistant and methicillin-susceptible and resistant clinical isolates of Staphylococcus aureus. Antimicrobial Agents and Chemotherapy 1998; 42:197-198.

Levy, S. B., *The Antibiotic Paradox: How Miracle Drugs Are Destroying the Miracle*. New York: Plenum Publishing, 1992.

Lin, Y. and Yuan. R. An ancient healing art in modern society. Bio/Pharma Quarterly 1997.

Nikaido, H. Outer Membrane. In: *Escherichia coli* and *Salmonella* Cellular and Molecular Biology. 2nd ed. Vol. 1. Neidhardt, F, Curtiss, R, Ingraham, J.L; Lin, E.C., Low, K.B., Magasanik, B, Reznikoff, W.S., Riley, M., Schaechter, M., Umbarger, H. E. eds. Washington, D.C.: ASM Press, 1996; 1:29-47.

Papageorgiou V.P., Assimopoulou A.N., Couladouros E.A., Hepworth D., and Nicolaou K.C. The chemistry and biology of alkannin, shikonin, and related napthazarin natural products. Angew. Chem. Int. Ed. 1999; 38:270-300.

Plyta Z.F., Li T., Mellidis A.S., Papgeorgiou V.P., Assimopoulou A.N., Couladouros E.A., and Pitsinos E.N. Inhibition of Topoisomerase I by Napthoquinone Derivatives. Bioorganic & Medical Chemistry Letters 1998; 8:3385-3390.

Sampson, B.A., Misra, R., and Benson, S.A. Identification and characterization of a new gene of Escherichia coli K-12 involved in outer membrane permeability. Genetics 1989; 122:491-501.

Skein, T., Kojima, K., Ota, S., Matsumato, T., Nagai, T. Preparation and evaluation of shikonin ointment for wound healing effectiveness in an experimental wound healing model in rats. S.T.P. Pharma Sci. 1998; 8:249-253.

Silhavy, T.J., Berman, M.L., and Enquist, L. *Experiments with Gene Fusions*, Cold Spring Harbor: Cold Spring Harbor Laboratory, 1984.

Tang, W and Eisenbrand, G. *Chinese Drugs of Plant Origin*. Berlin: Springer-Verlag, 1992.

Yoon, Y., Kim, Y. O., Lim, N.Y., Jeon, W.K., Sung, H.J. Shikonin, an ingredient of Lithospermum erythrorhizon induced apoptosis in HL60 human premyelocytic leukemia cell line. Planta. Med. 1999; 65: 532-535.

Yuan R. and. Lin. Y. Traditional Chinese medicine: an approach to scientific proof and clinical validation. Pharm. and Therapeutics 2000; 86:191-198.

ACKNOWLEDGEMENTS

This work was support by MIPS-grant #2502. Brian Higgins and Chi Chae were supported by HHMI undergraduate research fellowships through the College of Life Sciences.

Chapter 13

ON THE QUALITY ASSESSMENT OF CHINESE PATENT MEDICINE

PEISHAN XIE AND YUZHEN YAN
Guangzhou Institute for Drug Control, Guangzhou, China

Abstract: Chinese patent medicine (CPM), developed upon therapeutic theories of traditional Chinese medicine (TCM), is mostly composed of complicated formula. The quality standard of this kind of medicine , as a whole, cannot be used to evaluate its therapeutic value; it can only be used to control the quality of the final products. This is due to the fact that no single active constituent of any crude herbal drug in the medicinal product can be responsible for the entire therapeutic activity of such a complicated formula in the treatment of "Zheng" (roughly corresponds to "syndrome"). Therefore the quality standards of Chinese patent medicine are different from that of chemical pharmaceuticals upon which the therapeutic value and side effects of the product can be evaluated.

1. INTRODUCTION

The current standards of quality assessment for herbal drugs are set by identifying and quantifying the active constituent(s) by means of spectrophotometric and/or chromatographic methods in addition to other general tests reported in pharmacopoeia. In the year 2000 edition of Chinese Pharmacopoeia (ChP 2000), there are 602 monographs (including 992 of crude drugs and CPM) that contain thin layer chromatographic (TLC) identifications. There are 308 monographs include Assay items such as volumetric, spectrophotometric, high performance liquid chromatography (HPLC), gas chomatography (GC) and quantitative TLC (QTLC). The general Test items are stipulated in most of the monographs.

Yuan Lin (ed.), Drug Discovery and Traditional Chinese Medicine: Science, Regulatory and Globalization, 125-136.
©2001 Kluwer Academic Publishers. Printed in the Netherlands.

2. MONOGRAPHS IN THE PHARMACOPOEIA

The monograph of Huangqi (also called Radix Astrgalus, Milkvetch root) is one of the typical examples included in Chinese Pharmacopoeia 2000 Edition[1]

Milkvetch root is the dried root of *Astragalus memberanaceus* (Fisch.) Bge. Var. *mongolicus* (Bge.) Hsiao or *Astragalus membranaceus* (Fisch.) Bge. (Fam. Leguminosae). The drug is collected in spring and autumn, removed from rootlet and root stock, dried in the sun.

[Description] Cylindrical, some branched, upper part relatively thick, 30~90cm long, 1~3.5 cm in diameter. Externally pale brownish-yellow or pale brown, with irregular, longitudinal wrinkles or furrows. Texture hard and tenacious, uneasily broke, fracture highly fibrous and starchy, bark yellowish-white, wood pale yellow, with radiate striations and fissures, the center part of old root occasionally rotten-wood-shaped, blackish-brown or hollowed. Odour, weak; taste, slightly sweet and slightly bean-like on chewing.

[Identification] (1) Transverse section: Cork consisting of many rows of cells. Phelloderm of 3~5 rows of collenchymatous cells. Outer part of phloem rays often curved and fissured; fibers in bundles, walls thickened and lignified or slightly lignified, arranged alternately with sieve tube groups; stone cells sometimes visible near phelloderm. Cambium in a ring. Xylem vessels scattered singly or 2~3 aggregated in groups; wood fibers existing among vessels; stone cells singly or 2~4 in groups, sometimes visible in rays. Parenchymatous cells containing starch granules.

Powder: Yellowish-white. Fibers in bundles or scattered, 8~30μm in diameter, thick-walled, with longitudinal fissures on the surface, the primary walls often separated from the secondary walls, both ends often broken to tassel-like, or slightly truncated. Bodered-pitted vessels colourless or orange, bordered pits arranged closely. Stone cells occasionally visible, rounded, oblong or irregular, slightly thick-walled.

(2) To 3 g of the powder add 20 ml of methanol and heat under reflux on a water bath for 1 hour and filter. Apply the filtrate to a prepared neutral aluminum oxide column (100-120 mesh, 5 g, 10-15 cm in internal diameter) and elute with 100 ml of 40% methanol, collect the eluate. Evaporate the eluate on a water bath and extract with 2 quantities of 20 ml n-butanol saturated with water, combine the n-butanol solutions and wash with 2 quantities of 20 ml of water. Discard the water solution and evaporate the n-butanol solution to dryness on a water bath. Dissolve the residue in 0.5 ml of methanol as test solution. Dissolve Astragaloside IV CRS in methanol to produce a solution containing 1 mg per ml as the reference solution. Carry out the method for thin layer chromatography using silica gel G as the coating

substance and chloroform-methanol-water (13:7:2, lower layer) as the mobile phase. Apply separately to a plate 2 µl of each of the two solutions. After developing and removal of the plate, dry it in air, spray with 10% sulfuric acid in ethanol, and heat at 105°C for 5 minutes. A brown spot in the chromatogram obtained with test solution corresponds in position and reference solution. Examine under ultra-violet light (365 nm), the same orange-yellow fluorescent spots are shown in both chromatograms.

[Test] <u>Total ash</u> Not more than 5.0%
   <u>Acid-insoluble ash</u> Not more than 1.0%
   <u>The residue of pesticides</u> Total BHC not more than 0.2 ppm;
            Total DDT not more than 0.2 ppm;
            PCNB not more than 0.1 ppm

[Extractive] Carry out the cold extraction method as described under the determination of water-soluble extractives, not less than 17.0%

[Assay] Take 1.5 g of pulverized sample, weighed accurately, to a Soxhlet's extractor, add 40 ml of methanol, soak overnight, and further a quantity of methanol, heat under reflux on a water bath for 4 hours, remove the methanol from the extract and concentrate to dryness, dissolve the residue in 10 ml of warm water, extract with 3 quantities, each of 20 ml, of butanol saturated with water, combine the butanolic extracts, extract with ammonia solution twice (20, 20 ml), discard the ammonia solution, evaporate the butanolic solution to dryness, dissolve the residue in 3~5 ml of water, cool down, elute with 50 ml of water through a column packed with D101 coarse bore resin (1.5 cm inner diameter, 12 cm length), discard the water eluate, elute successively with 30 ml of 40%ethanol and 70% ethanol, discard the 40% ethanolic eluate, collect the 70% of ethanolic eluate and evaporate to dryness. Dissolve the residue in methanol and transfer to a 2 ml of volumetric flask, add methanol to the volume, and shake well, used as the test solution. Dissolve Astragaloside IV CRS, weighed accurately, in methanol to produce a solution 1 mg per ml as the reference solution. Carry out the method for thin layer chromatography (Appendix),. Apply 2 µl and 6µl of the test solution and 2µl and 4µl of the reference solution on the same Silica gel G thin layer plate, using Chloroform-methanol-water (13:6:2; lower phase after staying over night at 10°C) as mobile phase, removal of the plate after developing, evaporate the remnant of the solvent, spray 10% sulfuric acid ethanolic solution, heat under 100°C, scan the chromatogram with λ_S=530 nm, λ_R=700 nm, measure the integrate value and calculate the content. .It contains not less than 0.040% of astragaloside IV ($C_{41}H_{68}O_{14}$).

128

[Processing] *Radix Astragali* Eliminate foreign matter, grade according size, wash clean, soften thoroughly, cut into thick slices and dry.
Radix Astragali (processed with honey) Stir-frying the slices of Radix Astragali as described under the method for stir-frying with honey until no more sticky to fingers.

[Action] To reinforce *qi* and strengthen the superficial resistance, and to promote the discharge of pus and the growth of new tissue.
Radix Astragali (processed with honey): To reinforce *qi* and invigorate the function of the *spleen*.

[Indications] Deficiency of *qi* with lack of strength, anorexia and stools, sinking of the *spleen qi* manifested by chronic diarrhea, prolapse of the rectum, hematochezi and abnormal uterine bleeding; spontaneous sweating due to weakened superficial resistance; edma due to deficiency of *qi*; abscess difficult to burst or heal; anemia; diabetes caused by internal *heat*; albuminuria in chronic nephritis; diabetes mellitus.
Radix Astragali (processed with honey): Deficiency of *qi* with lack of strength, anorexia and loose stools.

[Usage and dosage] 9~30 g.

[Storage] Preserve in a ventilated dry place, protected from moisture and moth.

It is obviously that the specification of Chinese Herbal medicine like this is modeled after the standards used for synthetic chemical pharmaceuticals. This approach is also applied in the Indian, British and German herbal pharmacopoeia. As for chemical pharmaceuticals, it is clear that there is a direct relationship between content and efficiency; purity and safety. However, from the viewpoint of TCM theory and practice, there is no single constituent in a composite formulation that could be relied upon to evaluate the overall therapeutic value of the formula. For example, there is no evidence for the direct relationship between the content of astragaloside IV – the test target and the efficacy of the *radix Astragali*. Needless to say the compound formulae are obviously much more complicated than the single herbal drug.

The non-linear feature of the theory and practice of TCM requires some kind of comprehensive evaluation to assess the quality of Chinese herbal medication. Among various analytical practices, chromatographic fingerprint identification is an approach for monitoring the quality consistency of herbal medication[3~9]. The examples herewith enclosed depict the feasibility of chromatographic fingerprinting application to quality assessment of CPM[10~14].

3. EXAMPLES OF USING THESE TECHNIQUES ON QUALITY CONTROL OF CHINESE HERBAL MEDICINE

3.1 GINSENG

The specification of ginseng in ChP 2000 includes microscopic inspection, TLC identification (comparison with the authentic sample) and the content limitation of ginsenosides Re+Rg1 (not less than 0.25%) by HPLC. It can be used for the authenticity assessment of the species and the basic quality requirement. However the planar chromatographic fingerprint not only reveals the authenticity but also provides further detail information related to the quality assessment. The differentiation among species to species, parts to parts of the same species can be expected, even the degradation of the ginseng saponins in some of ginseng compound formula can also be traceable (Figs.1-4).

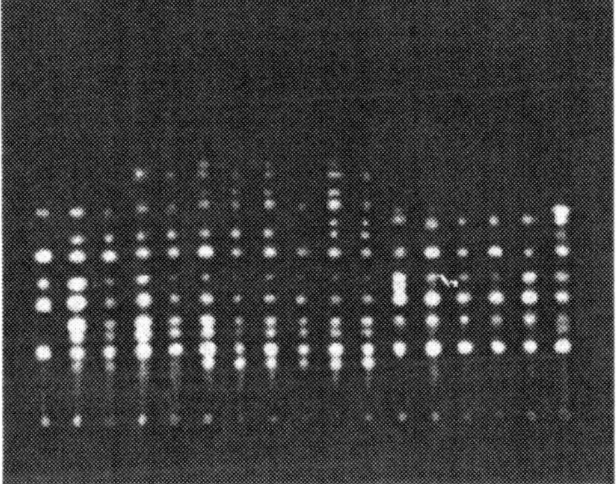

Figure 1. Fluorescent TLC images of different species of ginseng

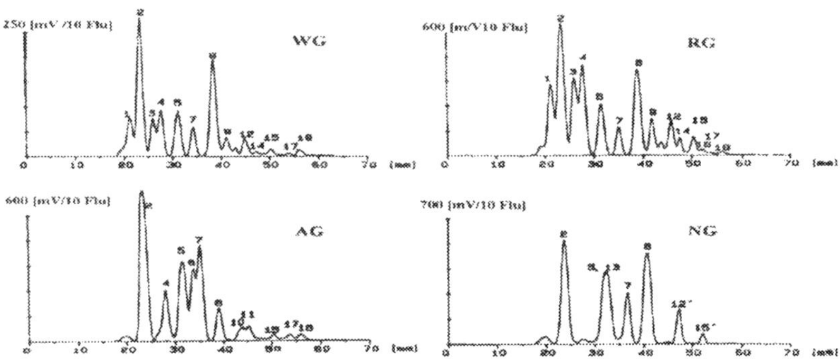

Fig. 2. Scanning profiles of TLC image of white ginseng (WG), red ginseng (RG), American ginseng (AG) and notoginseng (NG)

130

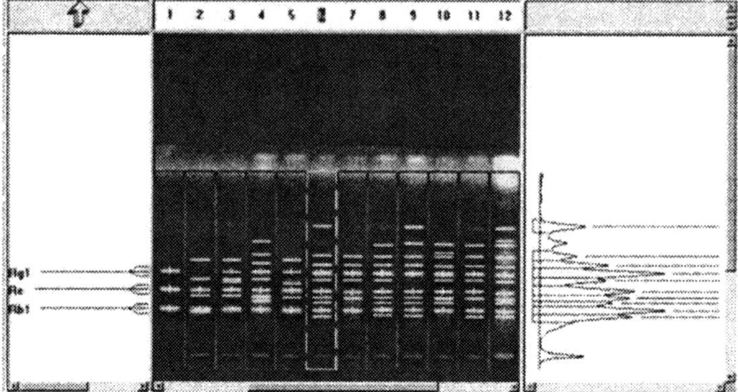

Fig. 3. Video-processing of TLC image of ginseng, American ginseng and notoginseng

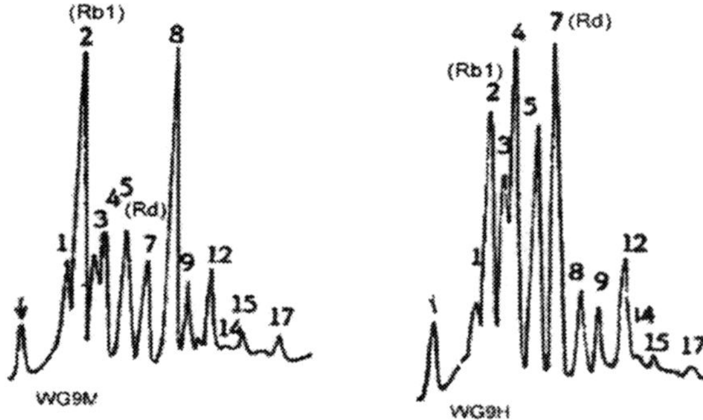

Fig. 4. Scanning profiles of the main root (left) and the root hair (right) of Ginseng. The ratio of Rb1 and Rd is significantly different between them.

3.1 RHIZOMA COPTIDIS (RHIZOME OF GOLDEN THREAD)

There are three species of *Coptis* that are used as Chinese herbal medicine under the name of 'Huanglian': *Coptis chinesis* Franch. (Chinese name: Weilian), *Coptis teeta* Wall (Chinese name: Yunlian) and *Coptis deltoidea* C.Y. Cheng et Hsiao (Chinese name: Yalian). It is impossible to distinguish these three species by using the main alkaloid (berberine) as reference substance . The planar chromatographic fingerprint can distinguish not only the common characteristic of the three species' chromatograms and also the 'minor' but significant difference among them (Fig. 5-7).

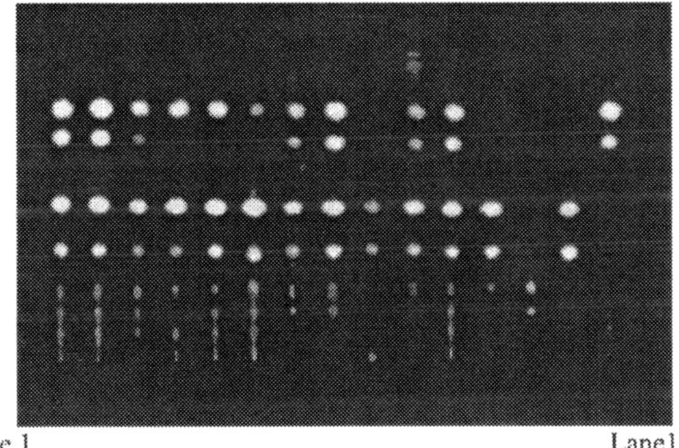

Fig. 5. Fluorescent TLC image of rhizoma Coptidis. Lanes 1,2. Coptis chineses; 3. C. deltoids; 4. C. teeta; 5: Yelian; 5: C. omeiensis;
7: C. C. japonica; 8,10,11: Patent CTM preparations containg Rhizome of Coptis spp.; 8.Patent CTM prep. Containing Cortex Phellodendri; 12~15:reference substances (columbamine, jatrorizine, palmatine, berberine, epiberberine+coptisine)

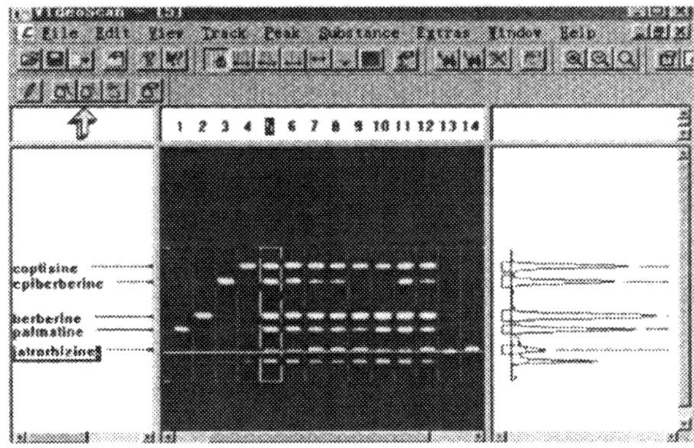

Fig. 6. Video-processing TLC images of rhizoma Coptidis

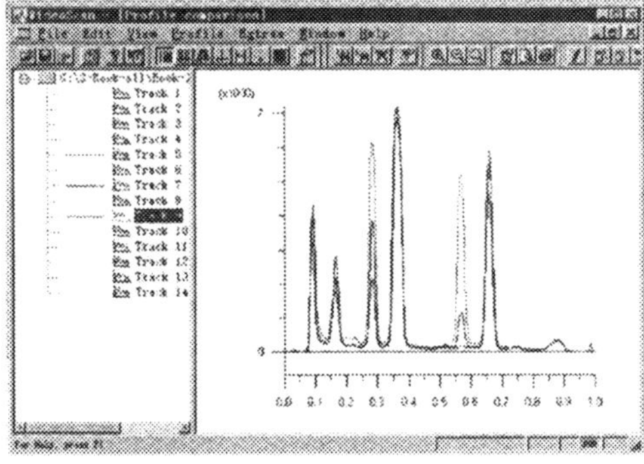

Fig. 7. Profile comparison of video-processing TLC image of Rhizoma Coptidis

2. 3 *GINKGO BILOBA* LEAVES

The monograph of *Ginkgo biloba* leaves and its standardized extract prescribes the content of total flavonoids and the total terpenes. But the HPLC fingerprint of ginkgo flavonoids is a powerful tool for evaluating the consistency of the products batch-by-batch and producer-by-producer (Fig. 8).

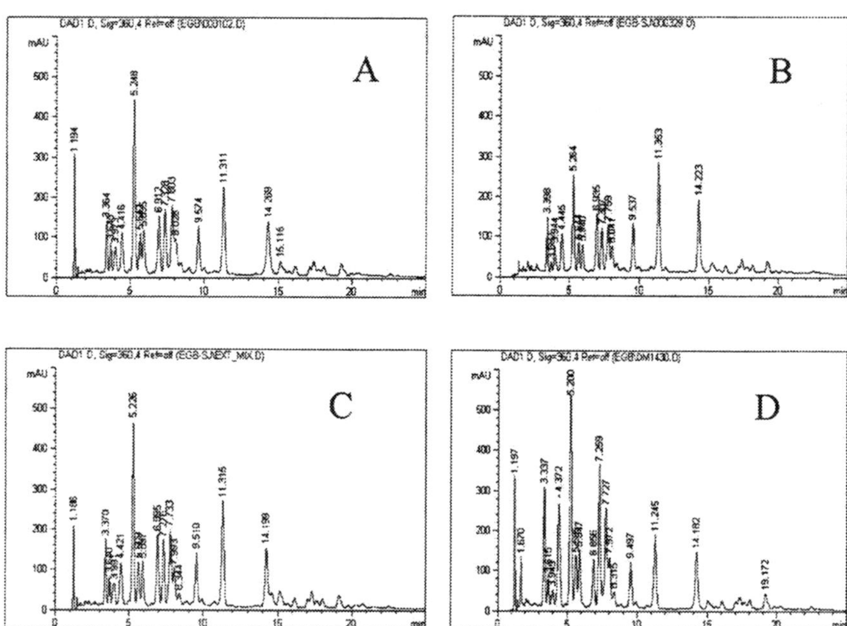

Fig. 8. HPLC fingerprint of Ginkgo flavonoides. A, B: commercial samples of Ginkgo preparations; C: a mixture of 8 batches raw materials of Ginkgo; D: Standardized Ginkgo extract powder (Egb761)

3.4 RHIZOMA *ALPINIAE OFFICINARUM* (LESSER GALANGAL RHIZOME)

For this species, only microscopical inspection of the rhizome is stipulated in ChP 2000. The chemical analysis published elsewhere only assay the content of 1,8-cineol by GC determination. The capillary GC chromatogram provided a consistent characteristic fingerprint of this species. It is easily to differentiate the authentic sample from the adulterations (Fig. 9).

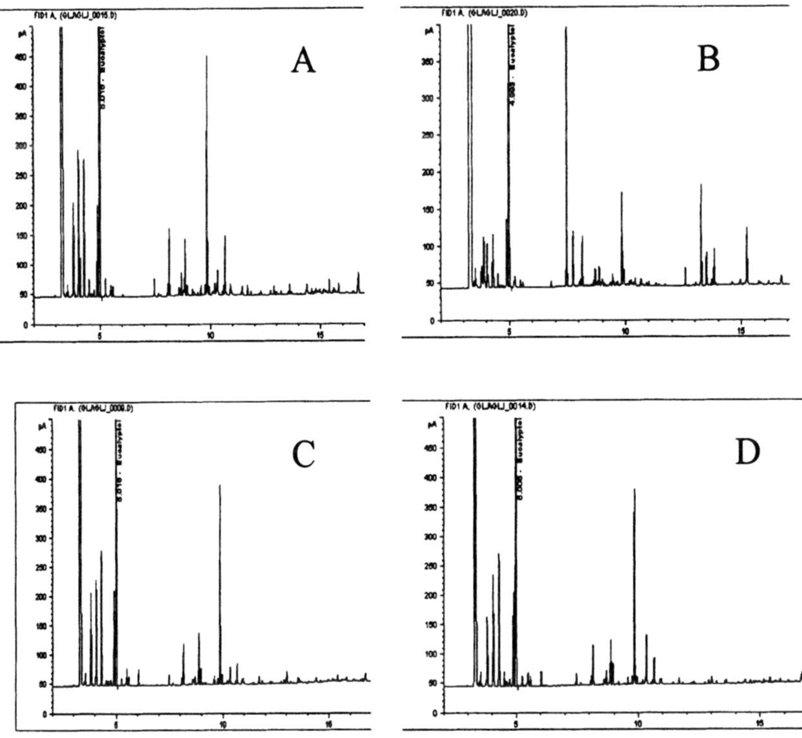

Fig. 9. GC fingerprint of Rhizoma Alpiniae. A, B, C: Rhizoma Alpiniae;
D a sample of adulteration Extracts

3.5 A PRIMARY INVESTIGATION OF HPLC FINGERPRINT OF A COMPOUND FORMULA

HPLC profile of a traditional Chinese patent medication – Xiao Qing Long Tang exemplified the capacity of fingerprinting analysis for quality assessment of multiple components sample by comparative studying the commercial products with the authentic reference sample (Fig. 10).

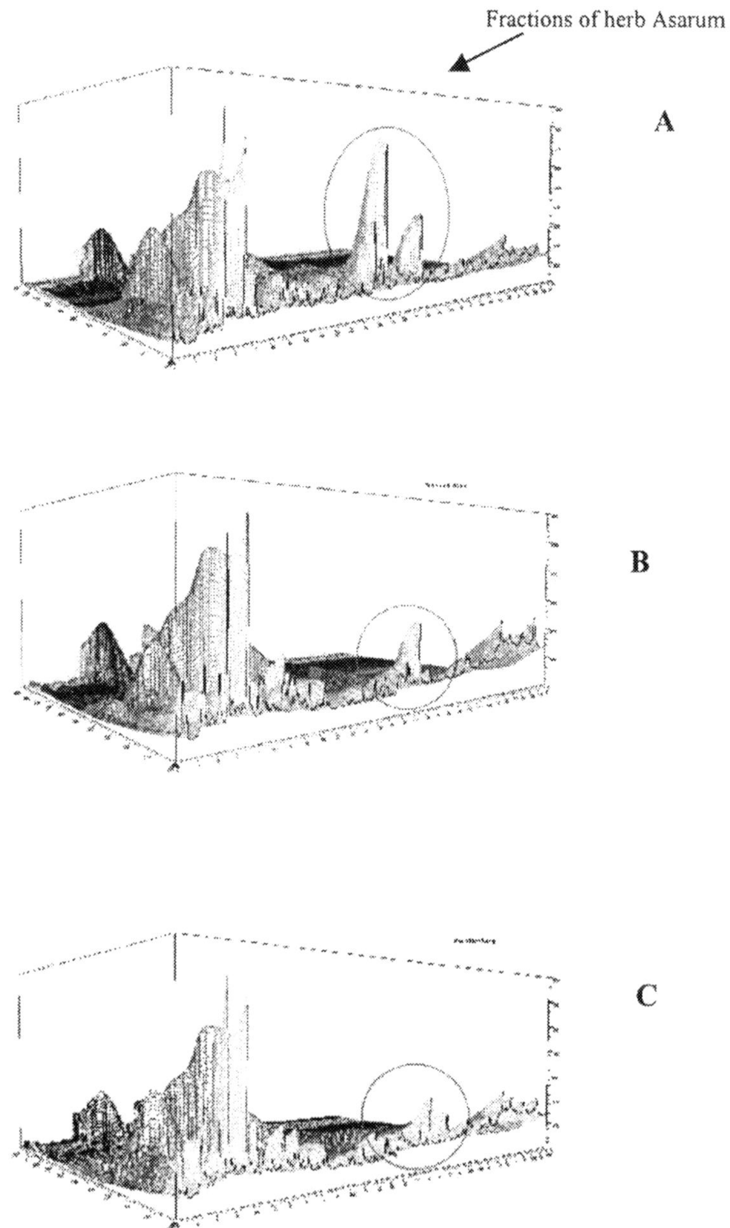

Fig. 10. 3-D HPLC fingerprint of Chinese patent medicine-Xiao Qing Long Tang.
A. Standardized sample; B, C. Two lots of commercial samples.

4. CONCLUSION

The assessment of quality consistency and stability of herbal medicine needs a more comprehensive approach than the current quality control methods commonly used in Chinese Pharmacopoeia, western herbal Pharmacopoeia and monographs. It will be the trends that the analytical chemistry develops from "decompose", "micro" analysis (individual constituent) towards "marco", synthetic analysis. These analyses include characteristic ultra-violet spectrophotometric spectra and fingerprint chromatograms of TLC, HPLC and/or GC even capillary electrophoresis (EC). More effective combination of qualitative and quantitative analysis can also be expected aiming at highlighting the effectiveness. Transition from "control standard" to "evaluation standard" of CPM by combining the physical-chemical analysis and the pharmacological evaluation of "group ingredients" such as bioassay will probably be the trends of the development of the quality standards of Chinese patent medicine.

REFERENCE

1. Chinese Pharmacopoeia (Volume 1), 2000 p.249
2. Xie Peishan On Integration of Traditional Chinese Medicine Culture and The Scientific Quality Control, J. of Integration of Chinese Medicine and Western Medicine 1998, 18 (11) 645-647
3. Philipsom J. D. British Herbal Pharmacopoeia. British Herbal Medicine Association Publications. 1996 Forward
4. Indian Drug Manufacturere's Association. Indian Herbal Pharmacopoeia . Vedams Books International. 1998 Volume 1
5. Branka Barl. Quality Analysis and Standardized Extracts of Medicinal Herbs. PMAP Conference 1997 (from Internet)
6. Rudolf Bauer. Quality Criteria and Phytopharmaceuticals: Can Acceptable Drug Standards be achieved? Drug Information Journal. 1998, 32: 101-110
7. Natalie Lazarowych. Use of Fingerprinting and Marker Compounds for Identification and Standardization of Botanical Drugs: Strategies for Applying Pharmaceutical HPLC Analysis to Herbal Products. Drug Information Journal . 1998, 32: 497-512
8. Peter John Houghton. Establishing Identification Criteria for Botanicals. Drug Information Journal, 1998, 32: 461-469
9. WHO. Guidelines for the Assessment of Herbal Medicines 1996
10. Xie Peishan, Yan Yuzhen. HPTLC Fingerprint Identification of Ginseng, High Resolution Chromatography & Chromatography Communication, 1987, 10(11): 607~610
11. Xie Peishan, Yan Yuzhen. Differentiation & Evaluation of Commercial Ginseng & Their Products by Means of HPTLC Fingerprint Analysis, 1988, 1(1): 29~32
12. Xie Peishan, Yan Yuzhen.. Application of HPTLC Fingerprint Analysis to Stability Evaluation of Ginseng Preparations, Journal of Planar Chromatography- Modern TLC. 1988, 1(3): 258

13. Xie Peishan, Yan Yuzhen. Optimization of the TLC of Protoberberine Alkaloids and Fingerprint valuation of the Coptidis Rhizome, Journal of Planar Chromatography-Modern TLC. 1992, 5 (5): 302-307

14. O. Sticher. Quality of Ginkgo preparations. Planta Medica . 1993, (59): 2-11

Chapter 14

MODERNIZATION OF TRADITIONAL CHINESE MEDICINE NEEDS FIVE FINGER MOUNTAIN AND GOLDEN HEAD RING

PAUL PUI-HAY BUT
Department of Biology and Institute of Chinese Medicine
Chinese University of Hong Kong, Hong Kong

Abstract: Chinese Medicine is just like the Monkey King, Sun Wukong, portrayed in the classic storybook "Monkey King (Journey to the West)." It is very dynamic, very powerful, and very capricious. It is also very unconventional and far from controllable. In order to tame it, so that it is ready for worldwide application, it is necessary to modernize it. The Five Finger Mountain and Golden Head Ring will do the job. The five fingers represent five areas: (a) authentication, (b) quality control from Good Agricultural Practice (GAP) to Good Manufacturing Practice (GMP), (c) safety documentation, (d) efficacy confirmation through evidence-based clinical and laboratory studies, and (e) computer informatics. Progress made in these five finger-areas would help to modernize Chinese Medicine. On the other hand, the Golden Head Ring, in the form of such requirements as GMP and other government policies and regulations, is helpful to keep ill-disciplined manufacturers and traders in order. Moreover, to help accelerate the modernization process, government funding agencies and the industrial sector should, like the Goddess of Mercy, provide more financial support. Once modernized, Chinese medicine would, like the Monkey King, achieve nirvana and obtain approval from the Leiyin Temple (e.g., FDA) in the West.

1. INTRODUCTION

Modernization of Traditional Chinese Medicine (TCM) has become an international subject (Yuan & Lin, 2000). Use of TCM is no longer limited to the Orient. There is an increasing consumer demand for TCM in the West

Yuan Lin (ed.), Drug Discovery and Traditional Chinese
Medicine: Science, Regulatory and Globalization, 137-144.
©2001 Kluwer Academic Publishers. Printed in the Netherlands.

138

(Eisenberg et al., 1998). More and more scientific studies have offered support to the values of TCM (Sheehan et al., 1992; Sheehan & Atherton, 1992; Bensoussan et al., 1998). However, regretfully, at the same time, more and more cases of adverse reactions to TCM have been revealed (But, 1994; But & Kan, 1995; Ko, 1999; Tomlinson et al., 2000; Fugh-Berman, 2000). Therefore, it is crucial to develop a strategy for the modernization of TCM.

The Chinese authorities have set the targets in a report on the strategies for the modernization of TCM (Gan et al., 1998). It advocated 'three low, three efficacious, and five convenient'. 'Three low' refers to 'low in dose, low in toxicity, and low in herb numbers'. 'Three efficacious' specifies 'strong efficacy, fast efficacy, and long efficacy'. 'Five convenient' demands for 'convenience for application, production, delivery, carrying, and storage'. Various researchers have suggested other models, such as the '6S Process' (selection, sourcing, structure, standardization, safety, and substantiation) (Chang, 2000). These targets and procedures emphasize essential elements in the modernization process. However, we believe that, at least among the Chinese people, it would be helpful to have an operation model that would create a more internalized incentive in the modernization task. Our proposed model follows the Chinese classical story "Monkey King," which is also known as "Journey to the West".

2. JOURNEY TO THE WEST AS AN OPERATIONAL MODEL

"Journey to the West" is a story about a legendary Monkey King that evolved into a very dynamic, very powerful and very capricious creature. The gods wished to tame it and keep it under control but never had success. The ghosts made every effort to trap it in hell, but in vain. Eventually, this Monkey King met the Absolute and boasted that he could easily stay outside the sovereignty of the Absolute. He made one somersault, flying some 180,000 miles away from the Absolute. There, he found himself landed on a mountain with five peaks. So, he wrote on the cliff of one of the peaks that he had taken a trip to that place. When he somersaulted back to the Absolute, he found his handwritings on a finger of the Absolute. To punish his arrogance, the Absolute placed the Monkey King under a Five Finger Mountain. There, the Monkey King suffered until an Expert came and removed all the taboos. Then, the Expert took the Monkey King along on a journey to the West. To avoid the Monkey King from temper tantrum and deviant actions, the Monkey King was given a golden head ring around his forehead. The golden head ring would contract and cause pain to the Monkey King, whenever the Expert read aloud certain prayers. After much turmoil and adverse situations, they arrived at Leiyin Temple in the West, obtained endorsement from the Temple, and eventually achieved nirvana.

TCM is very much like the Monkey King. It is very dynamic, very powerful, and very flexible. For example, different proportional combinations of a mixture of herbs would cover different health parameters to suit each individual health condition. As compared to modern medicine, it is also very unconventional and far from controllable. Modernization of TCM would thus need the Five Finger Mountain and Golden Head Ring mentioned in the story. The five fingers represent five areas: (a) authentication, (b) quality control starting from GAP to total chemical and functional consistence, (c) safety documentation, (d) evidence-based confirmation of efficacy, and (e) computer informatics. Progress made in these areas would help to modernize TCM. The Golden Head Ring, in form of such requirements as GMP and other government policies and regulations, would keep ill-disciplined manufacturers and traders in line. Like the Goddess of Mercy, government agencies should also offer positive reinforcement for good jobs done by researchers and manufacturers. Likewise, the industrial sector should mobilize more support for scientific research and development. Once modernized, Chinese medicine would, like the Monkey King, achieve nirvana and obtain approval from the Leiyin Temple (e.g., FDA) in the West.

3. ACTION IN THE CHINESE UNIVERSITY OF HONG

The Chinese University of Hong Kong (CUHK) established the Chinese Medicinal Material Research Centre (CMMRC) in the late 1970's for research in TCM. Despite receiving little support, the CMMRC maintained an impressive program in this area. Recently, CUHK elevated CMMRC to an Institute of Chinese Medicine (ICM). Five major sections were established to cover clinical trials, drug development, information, public and general education, and also standardization and safety. The most noteworthy change is the interest of many clinical colleagues to undertake or initiate evidence-based evaluation of TCM (Leung & Fung 2000; Critchley et al. 2000). A brief summary is given below to illustrate some of the efforts in harnessing the 'TCM Monkey King' undertaken by my research team in collaboration with various colleagues.

3.1 AUTHENTICATION

Hong Kong has served as a major entrepot for TCM since the latter parts of the Qing Dynasty. Over 80% of the American ginseng produced in North America first goes to Hong Kong for grading and then is redistributed to China and other places (But et al., 1995). Huge quantities of herbal materials are imported into Hong Kong every year for local consumption and re-export. There is a need for us to engage actively in undertaking research and offering

140

services in the authentication of TCM. A Museum of Chinese Medicine was established in 1984 and is now holding over 2,600 herb specimens. The museum is devoted to the documentation of TCM in the market and to the development of standards and protocols for the authentication of herbal materials. The standard pharmacognostic techniques use morphological (But & Ma, 1995; Mikage et al., 1987), anatomical (But et al., 1985; Ng et al., 1991) and chemical parameters (But et al., 1997; Chan et al., 2000). Moreover, we have pioneered applying molecular techniques to the authentication of dried Chinese herbal materials (Cheung et al. 1994; Ngan et al., 1999; Zhang et al., 1999) and have received a US patent for the application of such molecular methods (Wang et al., 1999).

3.2 QUALITY CONTROL

The quality consistence of TCM is a major concern. A laboratory for the chemical quality control of TCM products was created in 1995 with support from the Hong Kong Government Industry Department. Its mission is to promote the modernization of Chinese medicines relating to the areas of quality standardization and assurance. It has been extensively collaborating with government agencies and the herbal industry in target-oriented investigations (Lin et al., 1998; Lam & But, 1999; Song et al., 1999; Chan et al., 2000). For example, a recent study in collaboration with the Consumer Council on Ginkgo leaf products retailed in Hong Kong revealed that a majority of the products surveyed are below standard (Anonymous, 2000). Currently, the laboratory is expanding its scope into microarray systems for control on functional consistence of TCM.

3.3 SAFETY

The ICM and the Bureau of Drug Adverse Reactions at CUHK, has provided consultation and assistance in cases of suspected herbal poisoning. Requests for assistance have come from clinicians, hospitals, and government agencies, including the police and coroner's court. As a result, the majority of the mini-epidemics of herbal poisoning in Hong Kong were first identified by us and then reported to the Hospital Authority and the Department of Health, for announcement through the mass media. The major causes of herbal poisoning in Hong Kong are (a) wrong herb, (b) poor quality, particularly inadequate curing and poor manufacturing practices, (c) excessive dosage, (d) wrong prescription or bad judgment of herbal practitioners, (e) prolonged usage, (f) interaction with Western medicine, (g) intentional contamination with Western medicines, (h) individual errors, and (I) individual idiosyncratic responses (But, 1994, 1997; But et al., 1994, 1996a,b; Tai 1992a, 1992b).

3.4 EFFICACY

Colleagues in CUHK have undertaken many programs and projects to evaluate and document the efficacy of TCM. Recently, clinical colleagues have initiated evidence-based clinical studies in various areas. My research team has undertaken some studies on the pharmacological properties and mechanisms of TCM and on the clinical assessment of various herbal products, covering such areas as fertility regulation (Kong et al., 1985), HIV (Xu et al., 2000), antioxidants (He et al., 2000), vasorelaxant (Xie et al., 2000), eczema (Fung et al., 1999), and endangered animals (But, 1995; But et al., 1990; Li et al., 1995). Current attention is focused on TCM for treating viral infection, cough, and gynecopathy.

3.5 COMPUTER INFORMATICS

An English database on TCM was established at CUHK in 1980 with partial support from IBM. This database was later transferred to an open system for easy access through the internet (Kan et al., 1996). This task has led to the work on standardization of medical subject headings for TCM informatics (But et al., 1997). We have further developed databases for TCM education and herbal poisoning (Siu et al., 1996).

4. CONCLUSION

TCM is a treasure from the Chinese civilization and now benefits of TCM can be offered to the West. This 'journey to the West' will not be easy, but definitely challenging and potentially rewarding. The five areas of authentication, quality control, safety assessment, efficacy documentation, and computer informatics under the Five Finger Mountain would help accelerate the modernization of TCM.

CUHK has contributed some preliminary work in the research and development of TCM. We sincerely hope that more experts would join us to remove the 'taboos' and also look forward to opportunities for collaboration in research and development in order to take TCM to the absolute nirvana.

ACKNOWLEDGEMENTS

Partial support was received from Innovation and Technology Fund (AF/181/97, AF/281/97).

REFERENCES

Anonymous. Test casts doubt on clinical benefits of Ginkgo leaf products with non-standardized extract. Choice 2000; 289:20-27.

Bensoussan A., Talley N.J., Hing M., Menzies R., Guo A., Ngu M. Treatment of irritable bowel syndrome with Chinese herbal medicine: a randomized controlled trial. JAMA; 1998; 280:1585–9.

But P.P.H. Herbal poisoning caused by adulterants or erroneous substitutes. J Trop Med Hyg 1994; 97:371–4.

But P.P.H. Save the rhinos? Save the bears? Yes…But what about my child lying sick in Bed! Abstracts Chin Med 1995; 6:123-127.

But P.P.H. Trouble-shooting in the cases of herbal poisoning analyzed by the Chinese Medicinal Material Research Centre. Proceedings of the International Symposium on Chinese Herbs, Pharmaceutical Society of Taiwan, Taipei, 7-8 June 1997, pp. 55-61.

But P.P.H., Cheng L., Kwok I.M.Y. Instant methods to spot-check poisonous podophyllum root in herb samples of clematis root. Vet Hum Tox 1997a; 39:366.

But P.P.H., Kan W.K. Adverse reactions to Chinese medicines in Hong Kong. Abstracts Chin Med 1995; 6:104–22.

But P.P.H., Kan W.K., Siu Y.T. *First Report on Research and Development of Standard Medical Subject Headings for Traditional Chinese Medicine.* CMMRC, Hong Kong, 1997b; 1-83.

But P.P.H., Li T.Y.S., But A.Y.K. Ginseng in Hong Kong. In: Bailey WG, Whitehead C, Proctor JTA, Kyle JT (Eds.) *The Challenges of the 21st Century*, Proceedings of the International Ginseng Conference, Vancouver 1994, Burnaby, B.C., Simon Fraser University, 1995; 44-49.

But P.P.H., Lung L.C., Tam Y.K. Ethnopharmacology of rhinoceros horn. I: Antipyretic effects of rhinoceros horn and other animal horns. J Ethnopharm 1990; 30:157-168.

But P.P.H., Ma P. Pharmacognostical study of a counterfeit Cordyceps retailed in Hon Kong. J Chin Med (Taipei) 1995; 6:259-64.

But P.P.H., Ng G.K.H., Kong Y.C. Pharmacognostical Studies on the Related Articles of Kudingcha in Hong Kong Market. Jap J Pharmacog 1985; 9:46-51.

But P.P.H., Tai Y.T., Young K. Three fatal cases of herbal aconite poisoning. Vet Hum Tox 1994; 36:212-5

But P.P.H., Tomlinson B., Cheung K.O., Yong S.P., Szeto M.L., Lee C.K. Adulterants of herbal products can cause poisoning. Brit Med J 1996a; 313:117

But P.P.H., Tomlinson B., Lee K.L. Hepatitis related to the Chinese medicine Shou-wu-pian manufactured from Polygonum multiflorum. Vet Hum Tox 1996b; 38:280-2.

Chang J. Medicinal herbs: drugs or dietary supplements? Biochem Pharm 2000; 59:211-9.

Cheung K.S., Kwan H.S., But P.P.H., Shaw P.C. Pharmacognostical identification of American and Oriental ginseng roots by genomic fingerprinting using arbitrarily primed polymerase chain reaction. J Ethnopharm 1994; 42:67-69.

Critchley J.A., Zhang Y., Suthisisang C.C., Chan T.Y., Tomlinson B. Alternative therapies and medical science: designing clinical trials of alternative/complementary medicines--is evidence-based traditional Chinese medicine attainable? J Clin Pharm 2000; 40:462–7.

Eisenberg D.M., Davis R.B., Ettner S.L., Appel S., Wilkey S., Van Rompay M., Kessler R.C. Trends in alternative medicine use in the United States, 1990–1997: results of a follow-up national survey. JAMA 1998; 280:1569–75.

Fugh-Berman, A. Herb-drug interactions. Lancet 2000; 355:134-138.

Fung A.Y.P., Look P.C.N., Chong L.Y.C., But P.P.H., Wong E. A controlled trial of traditional Chinese herbal medicine in Chinese patients with recalcitrant atopic dermatitis. Int J Derm 1999; 38:387-392.

Gan S.J., Li Z.J., Z J.Q. Strategies for the Modernization of Chinese Medicine. Science & Technology Literature Publisher, Beijing, 1998.

He Z.D., Lau M.K., Xu H.X., Li P.C., But P.P.H. Antioxidant activity of phenylethanoid glycosides from Brandisia hancei. J Ethnopharm 2000; 71:483-486.

Kan W.K., Siu Y.Y., But P.P.H. Chinese medicinal databases at the Chinese University of Hong Kong. *The Hong Kong Library and Information Network - a Virtual Gateway to China.* Proceedings of the HKLA Conference, 21-23 Aug. 1996, Hong Kong. pp.71-75

Ko R.J. Causes, epidemiology, and clinical evaluation of suspected herbal poisoning. Tox Clin Tox 1999; 37:697–708.

Kong Y.C., Ng K.H., Wat K.H., Saxena K.F., Cheng K.F., But P.P.H. , Chang H.T. Yeuhchukene - a novel anti-implantation indole alkaloid from Murraya paniculata. Planta Med 1985; 304-307.

Lam K.W., But P.P.H. The content of zeaxanthin in Gou Qi Zi, a potential health benefit to improve visual acuity. Food Chem 1999; 67:173-6.

Leung P.C., Fung K.P. A practical comprehensive approach to Chinese medicine research. 1st International Conference on Traditional Chinese Medicine: Science, Regulation and Globalization, University of Maryland, College Park, Aug 30 – Sept 2, 2000.

Li Y.W., Zhu X.Y., But P.P.H., Yeung H.W. Ethnopharmacology of bear gall bladder: I. J Ethnopharm 1995; 47:27-31.

Lin G., Zhou K.Y., Zhao X.G., Wang Z.T., But P.P.H. Determination of pyrrolizidine alkaloids by on-line high performance liquid chromatography mass spectrometry with an electrospray interface. Rapid Com Mass Spectrum 1988; 12:1445-1456.

Mikage M., But P.P.H., Fong S.M., Namba T. Pharmacognostical studies on the Chinese crude drug "Ce bai ye" II. Jap J Pharmacog 1987; 41:161-168.

Ng T.H.K., Chan Y.W., Yu Y.L., Chang C.M., Ho H.C., Leung S.Y., But P.P.H. Encephalopathy and neuropathy following ingestion of a Chinese herbal broth containing podophyllin. J Neur Sci 1991; 101:107-113

Ngan F.N., Shaw P.C., But P.P.H., Wang J. Molecular authentication of Panax species. Phytochemistry 1999; 50:787-91.

Sheehan M.P., Atherton D.J. A controlled trial of traditional Chinese medicinal plants in widespread non-exudative atopic eczema. Brit J Derm 1992; 126:179–84.

Sheehan M.P., Rustin M.H.A., Atherton D.J., Buckley C., Harris D.J., Brostoff J., Ostlere L., Dawson A. Efficacy of traditional Chinese herbal therapy in adult atopic dermatitis. Lancet 1992; 340;13–7.

Siu Y.T., Kan W.K., But P.P.H. A CD-ROM for the identification of 100 potent Chinese herb. *Programs and Papers, International Symposium on Information and Publications of Chinese Medicine.* China Medical College, Taichung, 24-25 May 1997, pp. 210-218.

Song J.Z., Xu H.X., Tian S.J., But P.P.H. Determination of quinolizidine alkaloids in traditional Chinese herbal drugs by nonaqueous capillary electrophoresis. J Chrom A 1999; 857:303-11.

Tai Y.T., But P.P.H., Tomlinson B. Adverse effects from traditional Chinese medicine: A critical reappraisal. J Hong Kong Med Ass 1993; 45:197-201.

Tai Y.T., But P.P.H., Young K., Lau C.P. Cardiotoxicity after accidental herb-induced aconite poisoning. Lancet 1992a; 340:1254-56.

Tai Y.T., Lau C.P., But P.P.H., Fong P.C., Li J.P.S. Bidirectional tachycardia induced by herbal aconite poisoning. PACE 1992b; 15:831-839.

Tomlinson B., Chan T.Y.K., Chan J.C.N., Critchley J.A.J.H., But P.P.H. Toxicity of complementary therapies: an Eastern perspective. J Clin Pharm 2000; 40:451–6.

Xie Y.W., Ming D.S., Xu H.S., Dong H., But P.P.H. Vasorelaxing effects of Caesalpinia sappan: involvement of endogenous nitric oxide. Life Sci 2000; 67:1913-1918.

Xu H.X., Wan M., Dong H., But P.P.H., Foo L,Y. Inhibitory activity of flavonoids and tannins against HIV-1 protease. Bio Pharm Bull 2000; 23:1072-1076.

Yuan R., Lin Y. Traditional Chinese medicine: an approach to scientific proof and clinical validation. Pharmacol Ther 2000; 86:191–8.

Zhang Y.B., Ngan F.N., Wang Z.T., Ng T.B., But P.P.H., Shaw P.C., Wang J. Random primed polymerase chain reaction differentiates Codonopsis pilosula from different localities. Planta Med 1999; 65:157-60.

Chapter 15

A PRACTICAL COMPREHENSIVE APPROACH TO CHINESE MEDICINE RESEARCH

P. C. LEUNG [a,c] , K. P. FUNG [b,c]
[a]Department of Orthopaedics and Traumatology,
[b]Department of Biochemistry,
 [c]Institute of Chinese Medicine, The Chinese University of Hong Kong
Shatin, NT, Hong Kong

Abstract: Chinese Medicinal Research has the tradition of confinement to the laboratories where issues like safety, extraction, phytochemistry, pharmacology, active components and other concerns of authentication are explored and answers sought. The results thus obtained could be exciting, but quite often, they carry little clinical value.

Laboratory research would undoubtedly continue because of some obvious impressive results. Modern examples include productions from the plant periwinkle, and the antimalarial 'qinghao.'

While clinical studies were not scanty in journals published in China, they were found to be too frequently ignoring the principles of good clinical practice, and not many such studies were designed with evidence-based understanding.

A new innovative comprehensive approach is being recommended. The Chinese method, or drug to be tried, is selected on the basis of need (where modern medicine failed to offer perfect healing) and efficacy (where recorded experience has already proven its clinical effectiveness). Such efficacy driven clinical trials could be designed according to the strict evidence-based principles and started on volunteers or patients who give absolute consent for the trial. This approach ensures rapid progress, which goes side by side with other laboratory studies on authentication and extraction. Favorable clinical results will give sufficient confidence to proceed to drug development.

*Yuan Lin (ed.), Drug Discovery and Traditional Chinese
Medicine: Science, Regulatory and Globalization*, 145-149.
©2001 *Kluwer Academic Publishers. Printed in the Netherlands.*

1. INTRODUCTION

In spite of the great scientific achievements and technological promises in modern medicine, all practicing physicians and surgeons realized that plenty of problems and deficiencies still exist. While mastering the technologies of modern diagnostics and scientific management, there was no more than 15% of ability to remove a disease while the remaining, was just controlled. Control meant incomplete cure where problems were expected. More problems will be expected in long-termed measures, and chronicity is where deficiencies lie.

2. CHINESE MEDICINE COMPLEMENTS MODERN MEDICINE

Modern medicine has been built on a clear understanding of the structure, function and pathology of the human body, going from the gross to fine molecular states. When the fine details are well understood, the treatment details are well formulated, resulting in satisfactory control. However, when the details are not clear, treatment cannot be perfect. All diseases entering chronicity would provide un-explained manifestations which therefore challenge modern medicine treatment.

Chinese medicine, for some reasons, failed to get linked to structure and function. Treatment therefore could only be indirect and holistic – supporting the vitae organs to promote a spontaneous control. Chinese medicine, therefore, stresses on the holistic line and individual considerations which will supplement our deficiencies in modern medicine.

Given the proven successes of modern medicine, one tends to believe : had modern medicine not confine its approach to mainly direct and microscopic, its achievements could have been even more impressive. If modern practitioners could enlarge their armamentarium by going holistic and consider more individualized treatment when necessary, they could be more capable of solving problems and satisfy more patients.

Table 1. Value of TCM and its complementarity to modern medicine

- For Practitioners:
  Alternative subsidy like other complementary treatments
- For Patients:
  Satisfy the unsatisfied
- For Economy:
  Market driven, Health expenses, Health food, Advertising need
- For Medical Science:
  Supplement Deficiencies in Scientific Achievement

However, modern practitioners must be responsible. We could apply only those treatment methods that we believe are good clinical practices; not all those that have passed on from tradition. Not only do we need to prove that the preparations used are chemically and biologically safe, i.e. free from heavy metals, pesticides, aflotoxins and harmful microbiological organism, but to prove that they are clinically effective.

Research on traditional Chinese medicine in the past had been largely done at the laboratories, from safety studies, chemical analysis, extractions, pharmacokinetics to the modern technology of molecular biology. Very few of these complex studies led to clinical applications. A few successful examples are available like Senacot and Qin Hao in China and periwinkle in France (which gave both to vincristine) (Table 2).

Table 2. Successful TCM drug development using scientific approach

1) Qinhaosu (Artemisia annua L): Antimalarial
2) Leigong teng (Tripterygium wilfordü): Anti rheumatoid
3) Huperzine A (Huperizia serrata): Acetyleholin esterase inhibitor, Dementia
4) Phytoestrogens: Anticancer, anti coronary

Reports from China by J. Tang in 1997 on clinical trials are plentiful but a careful analysis of the 500,000 reports in 28 journals within the last 15 years indicated that there were problems with randomization and interpretation of results.

3. EVIDENCE BASED CLINICAL TRIALS

Perhaps there is a need to start from the very beginning of scientific clinical observations. There are areas of our need could identify while at the same time also identify highly effective or moderately effective treatment modalities in traditional Chinese medicine as means of satisfying our need. The matched problem and solution could be subjected to evidence-based clinical trials for confirmation and endorsement.

There are two approaches to clinical trials using traditional medicine. One is to work on basic sciences, find out the essential ingredients; improve the way of delivering these substances to the human body, work out the exact pharmoco dynamics, before applying to evidence-based clinical trials. Another way is to go straight ahead on evidence based clinical trials (Table 3) for those items that are urgently wanted as solution to difficult problems, provided that they have cleared the test on toxicity and safety.

The latter Efficacy Driven approach is preferred because it is faster. In fact a large number of laboratory studies were done in the laboratories leading no nearer to their clinical application. The WHO office on complimentary medicine has already endorsed this approach and we would like to follow the same line.

Table 3. Efficacy driven Approach

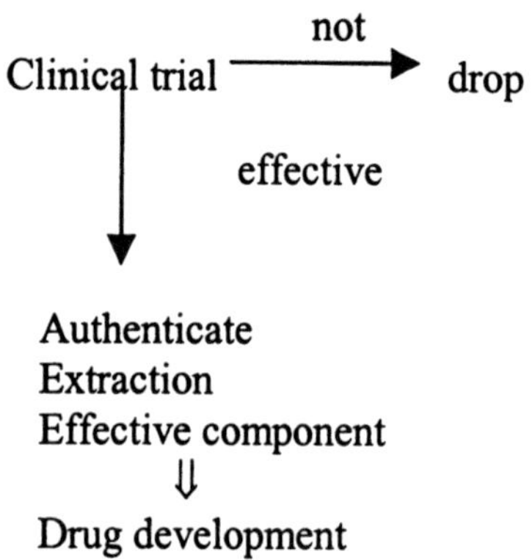

Where do our great needs lie? They are the areas that responsible modern practitioners also face difficulties while treating their patients. Common observations include: chronic pain, allergy, autoimmue problems, viral infections advanced cancers and psychosomatic. We need not be able to succeed getting the solution from traditional Chinese medicine, but should be well aware of the option and try.

The results of the evidence-based trial lead not only to popular clinical application, but also to the possibility of drug development, when pharmaceuticals would naturally get interested. Those that do not qualify after the evidenced-based trials would be automatically eliminated.

4. PROJECTS IN THE CHINESE UNIVERSITY OF HONG KONG

Under the Institute of Chinese Medicine of the Chinese University of Hong Kong; more than two dozens of evidence-based trials have been lined up. The projects that were already started included the following:

1. Limb salvage for diabetic ulcers using simple surgery and herbal supplement.
2. Reducing the need for steroid in asthmatic children.
3. Herbal remedy for terminal non-small cell cancer of lung.

4. Acupuncture treatment for enuresis.
5. Acupuncture treatment for chronic back pain not responding.
6. Herbal skin ointment for herpes.
7. Treating hepatitis B patients with Phyllantus.
8. Clinical trial on a common tonic for women – Dangque.

Projects that are being planned include the following:
Herbal adjuvant therapy to ensure a qualified cytotoxic drug level.

1. Herbal treatment for bone metastases.
2. Herbal treatment for myeloma.
3. Herbal treatment for limb swelling after injuries.
4. An eye tonic.
5. Acupuncture for the control of nausea and vomiting in pregnant woman
6. Acupuncture for the control of nausea in patients receiving cytotoxic drug.
7. Herbal antitussive.
8. Herbal mouth gargle.
9. Treatment for prostatic cancer.
10. Treatment for liver cancer.
11. Treatment for the hemiplegies and paraplegics.
12. Herbal remedy for osteoporosis.
13. Herbal remedy for senile dementia and Alzheimer's.

Since we launched our campaign on an objective look into our ability and deficiency in modern medicine, many colleagues in different departments indicated their need and support. We are confident that within months, all the projects will be put on line.

5. CONCLUSION

With so many decades of successes and wonderful achievements in modern medicine we should be certain by now, where our real strength lines and where is our Achillis heal, we should realize we won many victories where we are certain about the causes and changes. In those situations that we yet do not understand, we do not enjoy much success. As responsible practitioners we look forward to have those areas served efficiently using traditional alternatives.

Chapter 16

BACK TO NATURE:
THE ALTERNATIVE PARADIGM FOR DRUG DEVELOPMENT

JASJIT S. BINDRA, FRANK C. SCIAVOLINO, DAVID B. MACLEAN, PAUL A. ARMOND, AND PIERRE G. ETIENNE
Pfizer Global Research & Development, Groton, CT 06340

Abstract: Nature has traditionally been a fruitful harbinger of new medicines. However, despite the history of past successes, interest in natural products as a source of new drugs has withered over the past two decades. The trend in drug research has been towards combinatorial chemistry and high throughput screens, relying on receptor mechanisms to identify leads for further development. Following passage of Dietary Supplement Health and Education Act (DSHEA) legislation in 1994 and the resurgence of interest in herbals and the dietary supplements market, natural products are again attracting attention as a potential source of new medicines. These developments are triggering powerful changes on how botanical mixtures are developed and regulated.

1. INTRODUCTION

Natural products have been an important source of new medicines. By some accounts, nearly half of all pharmaceuticals have their origins in natural products (Figure 1) (Cragg, 1997). Morphine, cortisone, penicillin, paclitaxel, and lovastatin are only a few examples of prototypical molecules occurring in nature that have led to medicines, which are safe, effective and quality assured (Drews, 1998; Phillipson, 1997; DeSmet, 1997). The largely empirical screening methods that produced so many of the useful natural product-derived drugs discovered during the middle of the last century have now fallen into disfavor because they are highly labor intensive and increasingly unproductive.

Yuan Lin (ed.), Drug Discovery and Traditional Chinese Medicine: Science, Regulatory and Globalization, 151-156.
©2001 *Kluwer Academic Publishers. Printed in the Netherlands.*

152

More recently, the trend in drug research has been towards molecular design and combinatorial chemistry or the use of automated chemical technologies to synthesize huge libraries of compounds, which can be screened in ultra high-speed ways against novel biological targets. The revolutionary advances in molecular biology, genetics, robotics, and information technology have propelled drug research in this new direction.

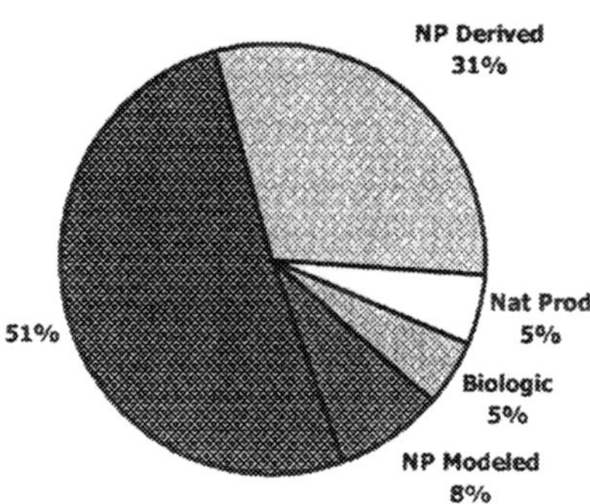

Figure 1: Nearly half of all pharmaceuticals have their origins in natural products (NP).

However, the link between pharmacological activity against newly discovered molecular targets and clinical effectiveness in specific disease indications is not always clearly understood nor fully charted, partly due to the inadequate predictive power of transgenic or animal disease models. Consequently, clinically validated mechanisms and clinically effective new chemical entities remain highly prized commodities and milestones in drug research and development.

2. NEW DRUG DISCOVERY

The road from an idea to a new medicine is long, treacherous, and expensive (Figure 2). This journey from a newly observed biological concept in the laboratory to a medically available registered drug typically takes 10-15 years, at a cost which has been estimated by the Tufts University Center for the Study of Drug Development to be in the range of $400-500 MM (DiMasi, 1995). Candidate attrition is a key factor contributing to the time and expense metrics associated with drug development. Most drug candidates nominated by sponsors for clinical development fail to become registered pharmaceuticals. For every five to six drug candidates that reach

Investigational New Drug (IND) status, industry experience indicates that only one becomes a product. The majority of drug candidates fail for a variety of technical reasons, most of which involve safety, efficacy, and the benefit-to-risk judgment. However, seemingly mundane pharmaceutical pitfalls, such as bioavailability and stability, may contribute to the decision to halt development. This is the reality of pharmaceutical R & D.

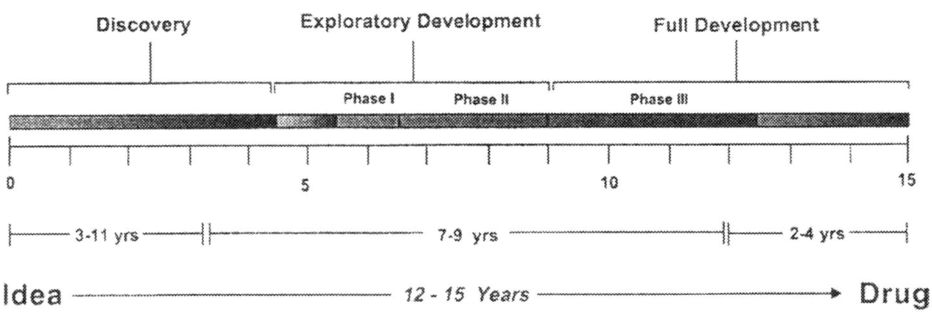

Figure 2: The new medicine timeline.

3. FROM DIETARY SUPPLEMENTS TO BOTANICAL DRUGS

With passage of the Dietary Supplement Health and Education Act (DSHEA) in 1994, a resurgence of interest in herbal medicines is taking hold, and natural products are again attracting attention as a potential source of new medicines. The increasing recognition of herbal medicine by regulatory agencies is a catalytic change that is kindling fresh interest in nature as a reservoir of useful medicines. Policy changes that may enable botanical drug regulation are beginning to emerge. The draft Guidance for Industry on Botanical Drug Products, which the FDA published for comment in August 2000, appears to be a meaningful first step toward opening the door for sponsors to conduct randomized, controlled clinical trials to confirm the efficacy of botanical mixtures in patients. This impending change could shift the focus of natural products research, from what was largely a laboratory effort, seeking lead structures for chemical modification, to a clinically driven initiative with product orientation. Final guidance from the FDA on the registration of botanical drugs would be a landmark event that could stimulate investment in this area by interested sponsors.

4. PFIZER'S APPROACH

In the natural medicines initiative at Pfizer, the term *naturaceutical* was used to designate prescription drugs of plant or natural origin that have been rigorously characterized for safety, efficacy, and quality. The development paradigm for *naturaceuticals* differs from the established pharmaceutical strategy in that it seeks up front to rapidly address clinical efficacy with candidates having anecdotal or folklore histories of use in humans, before investing in costly, time-consuming R&D work (Figure 3). Opportunities with proven clinical efficacy may then become fully invested for the costly process of registration of a safe and effective product. Products may be developed either as standardized, heterogeneous mixtures or as purified, single-chemical entities. While this approach appears to turn conventional R&D on its head, it only acknowledges the way drugs were discovered once upon a time. It is also a valuable mechanism for generating clinically validated leads.

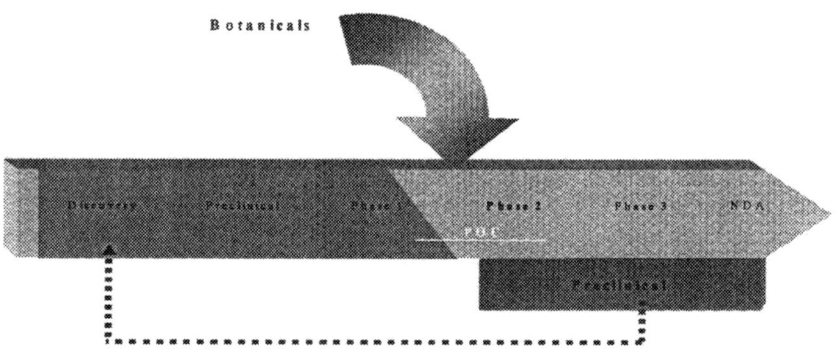

Figure 3: The naturaceutical paradigm – address proof-of-concept (POC) clinical efficacy upfront with candidates having anecdotal or folklore history of use in humans.

While emphasis during the last half century has been on single chemical entities, there has been a latent recognition that 'magic bullets' may be inadequate for combating diseases that are polymorphic in nature. Thus, the potential synergy of several active components in herbals and natural medicines may also confer a unique advantage (Figure 4).

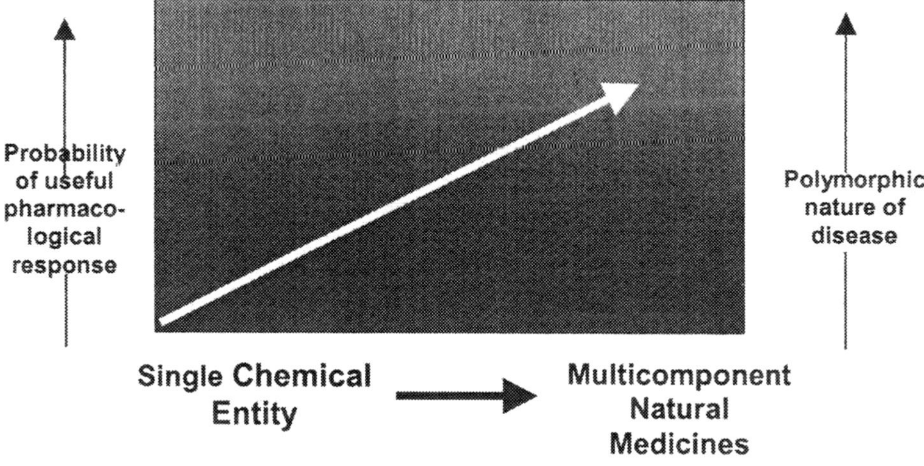

Figure 4: Single vs. multicomponent strategy against a complex biologic target.

5. CONCLUSIONS

If botanical medicines are to be developed as true pharmaceuticals, i.e., *naturaceuticals*, their development must be conducted to the highest standards of safety, efficacy, and quality. There can be no compromise on the science or medicine of *naturaceuticals*. The level of investment that would be required for the NDA registration of a *naturaceutical* is likely to be comparable to a pharmaceutical. Therefore, a key issue in developing botanical medicines is the need for an incentive mechanism to provide for a return on investment. One possibility is a Waxman-Hatch type data or marketing exclusivity period that would allow the sponsoring investor to fully develop the medical and commercial potential of natural medicines.

Nature has been an unparalleled source of unique chemical structures with extraordinarily potent pharmacological activity. We expect a supportive regulatory environment will be an important incentive for the private sector to invest more substantially in the research and development of natural medicines. Reminiscent of the early days of biologics, the current developments on the scientific and regulatory fronts with *naturaceuticals* are encouraging and may open the door to productive new therapies for humanity's many still unmet medical needs.

ACKNOWLEDGEMENT

Special thanks to Dr. Freddie Ann Hoffman for helpful discussions.

REFERENCES

Cragg, G. M., Newman, D. J., Snader, K. M., J. Nat. Prod. **60**, 52-60 (1997)

De Smet, P. A. G. M., Drugs, **54**, 802-840 (1997)

DiMasi, J. A., Hansen, R. W., Grabowski, H. G., Lasagna, L., Research and development costs for new drugs by therapeutic category, Pharmaco Econ, **7**, 152-169 (1995)

Drews, Jurgen, "In Quest of Tomorrow's Medicines", 23-48, Springer (1998)

"Guidance for Industry – Botanical Drug Products (Draft Guidance)" FDA Center for Drug Evaluation and Research, August 2000

Phillipson, D. J., Herbs, **22**, 17-19 (1997)

Chapter 17

GLOBAL MARKET FOR BOTANICAL PRODUCTS

PEGGY BREVOORT
2125 First Ave. Seattle, WA 98121

Abstract: The past few years have seen dramatic changes in the climate for botanical products both in the U. S. and throughout the world. The impressive growth of the market in the U.S. (fueled by passage of the Dietary Supplement Health and Education Act of 1994) has now leveled off leaving the industry with overstock of product and a somewhat uncertain future. On the other hand, the U.S. FDA has just published their draft guidelines for development of botanicals medicines as drugs, reflecting a growing interest in the drug model for these products. This chapter will cover a picture of the global market for botanical products as well as looking at the current U.S. market including market figures, best selling products and current surveys of consumer attitudes. It will also include a brief discussion of the TCM market and the potential for its development in the U. S.

1. INTRODUCTION

Even though the topic is the global market for botanicals, this chapter will be mostly focusing on the US market for botanicals. Because there is a lot more information on the US market and secondly, it is a good model to show ways that this new botanical industry is developing since the US is late to the game.

However, there are statistics on the size of the market worldwide for botanicals, an estimate of total market of 18 to 20 billion US dollars (Fig. 1). Asia, the largest market is at about 40%; Europe, 35%; North America, 17 %; Latin America 3%.

Yuan Lin (ed.), Drug Discovery and Traditional Chinese
Medicine: Science, Regulatory and Globalization, 157-168.
©2001 *Kluwer Academic Publishers. Printed in the Netherlands.*

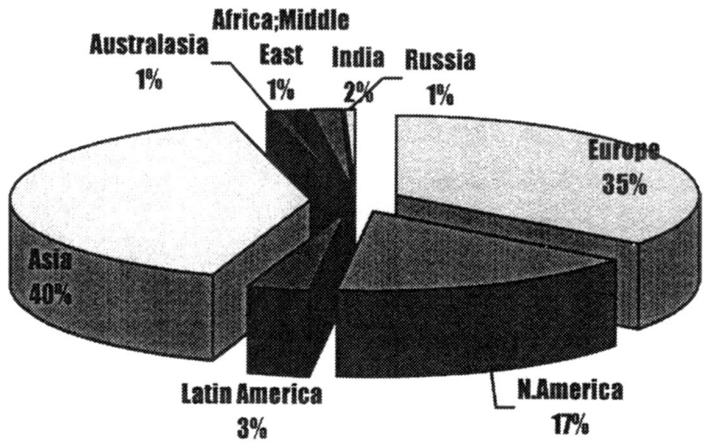

Fig. 1.Estimate world market for botanical medicine (1997) in US $billion

2. GLOBAL MARKET OF HERBAL PRODUCTS

Table 1.below lists issues that, in fact, not only affect the global market but also the U.S. market.

Table 1. Issues Affecting the Global Market

- Population of post world war II - "baby boomers"
- Escalating expense of western pharmaceuticals
- Emergence of "global economy" with ease of communication - e.g. Internet
- Desire to return to "natural lifestyle"
- Renewed interest in ethnic cultures and "roots"
- Growth of U. S. market as growth "trend setter"
- Interest in "Nutraceutical" products
- Drug companies with global presence now entering market (Bayer)

There are also ecological issues that are of great concern to the global market: the concerns about the long term supply of raw material; industrial developer and over harvest. Other issues such as genetically modified organisms; the rights of indigenous people are very complex subjects. The intellectual properly rights that are associated with drug development or new product development anywhere in the world, and the concern about our increasing industrial contamination of the entire world that might affect our initial material.

Today, the global market of herbal products is affected by the regional changes in regulatory or marketing strategies. Table 2 below illustrates some of the changes that affect the global market.

Table 2. Changes in global market

- European quality standards as world model
- U.S.A. - Impact of DSHEA felt around the world
  --Increased U.S. market
  --Model for regulations
- S. America looking for regulatory models
- Japan increasing # of herbs approved as foods
- China concerns with quality and export of TCM
- Africa development of export market
- India - Export of Aryuvedic products to world
- E-Commerce: The new country?

Certainly at this point, the European quality standards have been the world model. Europe has been involved in botanicals without a broken history, longer than the US. And they also applied a lot of industrialized standards to the products. The new legislation in the US has impacted the regulatory environment around the world. Japanese are increasing their use of herbs. China is concerned with quality and the export of their traditional Chinese medicine (TCM) products. Indeed, e-commerce maybe the fastest way for the globalization of this industry.

Fig. 2. Varying distributional channel for herbal products

3. CONSUMPTION OF HERBAL PRODUCTS IN THE U.S.

It's very difficult to track different marketing channels for herbal products around the world particularly in the multi-level marketing (Fig. 2). Everywhere there are small local store sales which could have missed in our tracking system. There are no records on the practitioner market for TCM; the E-commerce is uncharted. There are ethnical sales as well. This is a tremendous market whose size is not well understood. There is more work to be done to understand how deeply penetrating these herbs are throughout the world. It will be very significant to our regulators and to people who perhaps don't understand their value.

In different parts of the world there are different routes to the consumer. For example, in the US, almost half of the consumers are buying their herbs from the specially food stores. Whereas in Europe, there is a much large percentage going through pharmacy or drugstore because of different regulations. And then in Japan over 50% are direct sales or multi-level marketing. So depending on the culture, there is a different primary channel.

It is estimated that 32% of the US population is using herbs on a regular basis; in Germany, that is >60%; in China it is 80- 90% of the population. So indeed the US is certainly growing but still has a lot of room to grow. According to Dr. David Eisenburg, ~4% of the population was using herbs in 1991 in the US; in 1998, 30-35% is the generally agreed figure of the percentage of the US population using this product.

According to *Prevention Magazine*, 49% of the population who had used herbal products in the past 12 months and about 24% of the population used these products regularly. Most frequently used products are garlic, ginseng, ginkgo, St. John's wart and Echinacea

Why are they using them? Ensure good health is the main reason followed by improving energy and then preventing or treating colds and flues.

Why would consumers be using remedy instead of over the counter products (OTC) in this country? Mostly likely it is because they prefer more natural products and this is a trend around the world that people are looking for the natural alternative. These products are most frequently used for treating cancer, nutrition, and heart disease.

And how long will consumers use herbal remedies without results? This is a very interesting observation: 28% of consumers would continue to use the product for about a month without any perceptible results. In a sense that's very significant because it will take that long for herbal products to show some effect whereas in western drug it usually takes less than 20 minutes. But it is interesting that there is a consumer level of faith.

The following table summarizes the regular categories within the US through which botanical products were sold. Dietary supplements are the primary focus currently; traditional medicines, which is not yet established as a category and yet there has been much discussion in the US about the possibility or perhaps the need for that traditional medicine's category. OTC and new botanical drugs are under FDA's reviewing process.

Table 3. Possible Regulatory Categories of herbal products

- Conventional Food
  NLEA Health Claims
- Dietary Supplements
  (DSHEA, 1994)
- Traditional Medicines
  Category not yet established
- OTC Drugs
  Petitions pending from European American Phytomedicine Coalition for Valerian and Ginger
- New Rx or OTC Drugs (IND/NDA route)
  100 plus botanical formulas already submitted

Natural medicine herbs, vitamins and minerals are still a very small part of the whole medicine industry within the US when compared to generic drugs (64 billion in sales) and OTC (24 billions).

There are some botanicals that have been approved as OTC ingredients as shown in Table 4.

Table 4. Botanicals Approved As OTC Drug Ingredients

Herb	**Approved Use**
Aloe (Aloe ferox)	Stimulant laxative
Cascara sagrada	Stimulant laxative
Peppermint oil	Antitussive
Psyllium	Bulk laxative
Red pepper	Counter irritant
Senna	Stimulant laxative
Slippery elm	Demulcent
Witch hazel	Astringent

There are already over a hundred botanicals either individually or in formulas currently going through FDA's clinical trial process. The guidance document that FDA has just published is a very significant document that paves the way for a new route for botanicals that in some way is very appropriate for them. But even before the document was published, there have been many companies and individuals that have submitted the Investigation New Drug Applications (INDs) and these are in process. The day the first IND is approved it will be a changing point in the industry. Table 5 sums up the botanical IND's by their clinical specialty.

Table 5. Some Botanical IND's by clinical specialty (1990-98)

- Anti-inflammatory   3
- Anti-viral   12
- Cardio/renal   2
- Dermatological   19
- Drug abuse   2
- Endocrine/Metabolism   3
- Neuropharm   7
- Oncology   13
- Urogenital   2

Table 6 depicts the trends within the US market. These trends also reflect worldwide changes.

Table 6. U.S.Botanical Market Trends-1999

- Market:
    - Entry by major "big pharma" with resources
    - More company entries; inventory overstock
    - IRI statistics steady rather than growth
    - Business consolidation (merger and acquisition)
    - Competitive pricing and "commoditization"
    - Price considerations overriding quality
    - Industry looking for new "blockbuster"
- Health Care
    - Increased usage and prescription by health practitioners
    - Increasing reimbursement by insurance and HMO
    - Development of clinical trials specific to botanicals
    - Interactions between herbs and pharmaceuticals
    - Adverse event reporting mechanism needed

Major companies with large resources now coming into the market. More company entries are overly optimistic about the future of the market. The market is now quite flat. The industry is not growing as it was 2 years ago. There have been a lot of business consolidations. This is natural in a mature industry. There is great deal more competitive pricing in a very competitive market, thus, more difficult to enter. And often times, unfortunately, price considerations override quality. This is will become a serious issue. And again the industry is looking for the brand new product that is going to take them out of the competition. In the area of health care, there is an increased demand for prescription drugs that can be reimbursed by insurance HMO. Development of clinical trials of botanicals, thus, is growing every day.

And there are concerns about the interactions between herbs and pharmaceuticals. Today there is an increasingly usage of complimentary and alternative medicine. The second study by David Eisenburg shows an increasing usage, increasing dollars and increasing consumer willingness to spend their own hard earned dollars on complimentary and alternative medicines.

The National Institutes of Health, National Center for Complimentary and Alternative Medicines is now supporting studies on the use of traditional indigenous systems of medicine. They are establishing a national center and to add complimentary and alternative information into the curriculum of medical schools. This is very important because unless the doctors understand these areas of medicine, it will not grow.

There are many concerns regarding the product development (Table 7). There are companies entering into the market without an expertise in understanding how to work specifically with botanicals. The future supply of raw material and the question about intellectual property are issues need to be addressed. Whether to use a formula or single herb; how to validate what is in the herb, in particular if the active ingredients are not known.

Table 7. Issues in Product development

- Lack of herbal knowledge by major new players
- Importance of future supply of raw material
- Exploitation of indigenous peoples
- Understanding synergy of whole plant or complex formula
- Combination formula vs. single herb product
- Bioassay or other type of validation
- Powdered herb vs. extract

164

The growth of the herbal market in the US is slowing down (Fig. 3). By the end of 1999, the total sale is under 4.5 billion. There is no statistics yet for the year of the 2000 but it will not be the type of growth seen in the past.

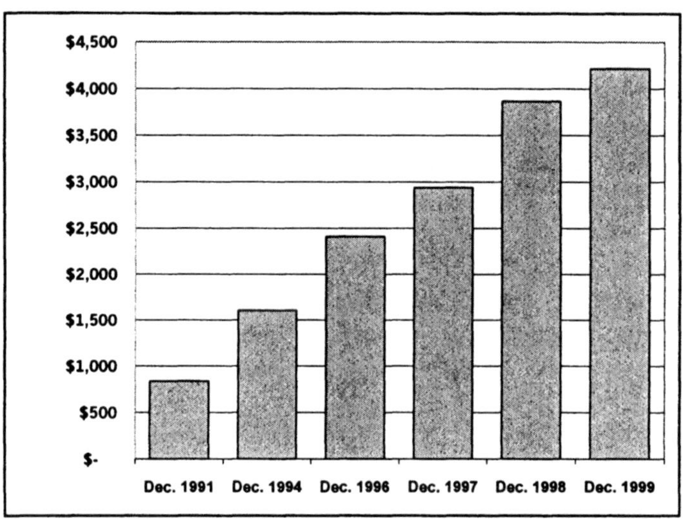

Figure 3. Growth of US market for botanical medicine 1991-1999

Mass market, such as the food, drug and mass merchandise like Wal-mart chains are the best track channels in the US (Fig. 4). And this was the fastest growing channel. A lot of companies decided that they could jump into the market because this tremendous growth in mass market to almost a billion alone by the end of 1998. Unfortunately or realistically, that market has now leveled off. This period of no growth could be one of the signals of the new change and in some ways indicative of how the industry will develop in the future.

It's interesting, however, that most new consumers are buying their herbal products at pharmacy, grocery, supermarket and not so much in the health food supermarket channel (Table 8). However, if one looks at what is considered to be the average price, in a multi-level health food store, the average price per unit of a purchase of a botanical is quite high, whereas in grocery and mass pharmacy there is a much lower average price (Table 9).

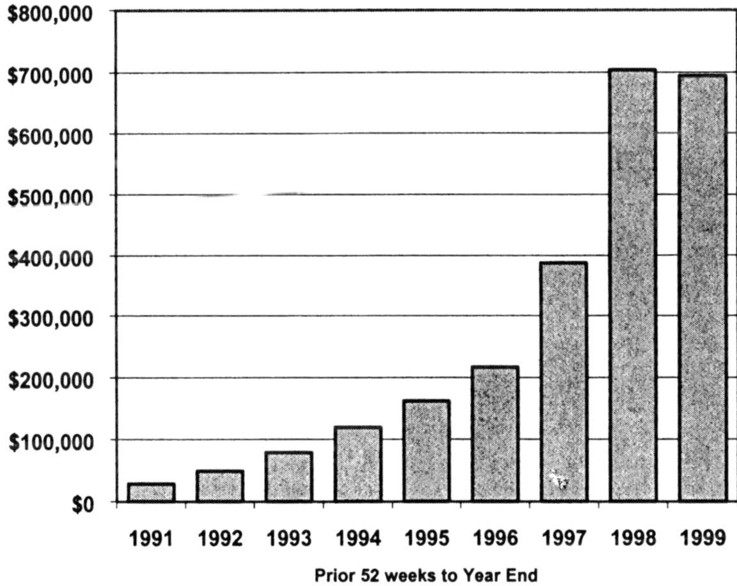

Figure 4. Mass market growth in the US 1991-1999

Table 8. Supplement Usage in the US (Hartman/New Hope Survey)

<u>Survey of 43,000 Household Purchase Patterns</u>

- Pharmacy/Drug   <u>32%</u>
- Grocery/Supermarket   <u>23%</u>
- Direct Mail   <u>17%</u>
- Health Food Store   <u>11%</u>
- Vitamin Store   <u>11%</u>
- Club Store   <u>8%</u>
- Health Food Supermarket   <u>6%</u>

- <u>Natural Foods Merchandiser Feb.'98</u>

This either means that there is a lot more cost cutting that due to efficiency or there are indeed corners been cut on quality.

Table 9. Herbal Product Average Retail Price by Market Channel
July'98-Apr.'99

Channel	**Average Price per unit**
Multi-level	$24.16
Health Food Store	$15.97
Health Practitioner	$15.39
Direct Mail	$14.09
"Pill Store"	$13.06
"Club Store"	$12.77
Pharmacy	$ 9.09
Grocery	$ 7.34
Mass Market	$ 7.12

FDC Reports/The Hartman Group: Dietary Supplement Market View, July 1999

Table 10 is the top selling herbal products in the US. Included in this table are Chinese herbs that are being sold more and more in many different channels.

Table 10. Top Herbal Ingredient Categories

Category	%of Sales	Avg.price	Total sales
• Garlic	9%	$ 6.82	$160M
• Ginkgo	9%	$10.31	$239M
• St.John's wort	7%	$ 8.72	$148M
• Echinacea	6%	$ 8.02	$109M
• Diet Supplements	6%	$18.94	$325M
• Ginseng	5.4%	$ 7.96	$121M
• Saw palmetto	3%	$10.87	$ 91M
• Energy Supplements	2.4%	$19.54	$114M
• Kava-kava	1.7%	$ 7.45	$ 31M
• Chinese herbs*(small sample) 1.5%		$10.31	$ 63M

FDC Dietary Supplement Market View, August,1999 pg 11.

4. TRADITIONAL CHINESE MEDICINE

The accurate information about the TCM market in the US is not easy to obtain. A private company which is also a selling Chinese medicine provide the following data. They surveyed across all economic levels and all types of occupations nationwide. According to their survey, Americans in general believe TCM works and it was the original herb medicine. Most people asked have positive attitude towards TCM.

Table 11. Survey of American Attitudes on Chinese Medicine*

- Americans think "it works"
- Surveyed across all economic levels
- Surveyed across all types of occupations
- Surveyed nationwide
- Americans believe Chinese medicine is "original herbal medicine"
- Many stories about success with Chinese herbs
- Positive attitude on Chinese herbs (good feeling)

*From proprietary industry survey, 1999.

Table 12 is an overview of TCM market in the US. Some of the product category is important particularly in practitioner community; these include products for stress, immune system, energy, digestion, and women's issues and clear thinking. The price range can go all the way from 4 dollars to >20 dollars per unit; and from a few cents to several dollars per dose, The high price end usually come from a practitioner, presumably a high quality product.

Table 12. TCM Market Overview-U.S.A.

Product Categories

- Stress
- Immune System
- Energy
- Digestion
- Women's issues
- Clear thinking

Price

- Range: US$ 4.25 - $23.95
- Per dose: US$ 0.23 - $1.41

Table 13 describes some of the TCM market facts; 12 million patients visit 7000 license practitioners yearly; 50 colleges of Oriental Medicine graduating thousands of graduates a year; there are 50 medical colleges now offering courses in complimentary and alternative medicine if not specifically on TCM.

Table 13. TCM: U.S. Market Facts

- 12 Million patient visits to 7,000 licensed practitioners
- 50 colleges of Oriental Medicine; Graduating 1,000 students per year total
- Practitioners who use TCM: Acupuncturist, O.M.D. Naturopathic, M.D.
- 50 Medical colleges offer some course in Complementary and Alternative Medicine.
- American Association of Acupuncture and Oriental Medicine

5. CONCLUSION

In conclusion, TCM has certainly entered the US market. Large sections of health food store are selling Chinese herbs. There are companies that are specialized in Chinese herbal products and making various formulations. Mass market, multi-level and drug stores are selling Chinese herbs.

For the year 2000 and beyond herbs are really a global market; Chinese herbs, American herbs, South America herbs, south pacific herbs will eventually be part of our daily lives. Scientists everywhere are going to use the very best methods to really understand these products. And as we moving to the 21st century, we must not forget that these were our roots and there were our medicines.

Chapter 18

OPPORTUNITIES AND CHALLENGES OF DEVELOPING MEDICINAL HERBS

MICHAEL CHANG
Pharmanex Inc. 2000 Sierra Point Pkwy, Brisbane, CA 94005

Abstract: There is a good opportunity for the development and marketing of TCM worldwide. Consumer demand for medicinal herbs is growing at a double-digit annual rate, and global regulatory environments are also changing rapidly. However, there are a number of hurdles that need to be overcome before TCM becomes part of mainstream heath care practices. This is Pharmanex's approach.

1. WORLD AND US MARKETS: TRENDS AND CHALLENGES

We use the US health industry to demonstrate the global market. Traditional Chinese medicine (TCM) has an opportunity in three areas to make a commercialized product. Clearly the largest one is the prescription drug market that is worth $90 billion in the US. Globally it's probably three times that, about $200 billion with a growing rate of 6% annually. The second category is the OTC drugs that do not require prescriptions from the doctors. The US market is about $26 billion, growing at about 4.7%. OTC drugs also have a very large global market of over $100 billion. TCM has a unique opportunity to participate in both categories. However, there's one key challenge: TCMs are indeed different from traditional, mainstream, synthetic, western drugs which doctors automatically prescribe, believing the regulatory agency screens the safety and efficacy of the product. TCM requires educated and competent professionals, even if it receives FDA approval. So there is a long way to go because there will be challenges in terms of marketing these products. The last category is herbals: dietary supplements and food, and this market is at $2.6 billion. It has been growing very rapidly showing double-digit figures for the past four or five years. However, in 1999 the growth has

*Yuan Lin (ed.), Drug Discovery and Traditional Chinese
Medicine: Science, Regulatory and Globalization*, 169-176.
©2001 *Kluwer Academic Publishers. Printed in the Netherlands.*

flattened out and there may even be a lessening in demand. The trend is also the same for TCM.

The largest group of users is disease prevention (Table1), even though with dietary supplements we're really not allowed to talk about medicating or preventing disease. People do use TCM to stay healthy by preventing the occurrence of disease, thus making doctors and prescription drugs unnecessary. The second largest category (greater than 50%) is for energy. And the following are for fitness, alertness, etc. These are the areas of optimal potential.

Table 1 shows conditions most commonly treated by supplement users.

Conditions	% users
Depression	21%
Medical Problems	25%
Stress	27%
Alertness	31%
Fitness	40%
Energy	54%
Disease Prevention	62%

Table 2 shows some of the most popular herbal products in the US market. There are three important points. First, if one looks at the top ten, the first couple has very high volume. Ginkgo alone is about $150 million in sales. That's number one in 1998 in the US. Globally, it is more than a billion dollars. St. John's Wort is number two at 140 million. But the sales of other products drop off very rapidly. Go down to number five or six, it's already down to a very small number. Number ten; valerian root is only $16 million. So not all herbs are selling very well in western market. The second point is that these are all single species herbs. Mainstream America is not quite comfortable about very complex formulations. Education is needed to build up confidence and credibility for TCM that is mostly multiple herbs. The third point is that although there are thousands of formulas, 70% of them are comprised of less than a handful of herbs. A pessimist might view this market as very hard to crack. An optimist would say that the potential is huge. If we were to get FDA's approval, we would have a regulatory method of really endorsing this product. We might also gain the confidence of professionals as well as the consumers.

Table 2. Top 10 Herbs in the US-1998 Ranking

Botanicals	Retail Sales 1998 (in million)	Change % 97-98
Ginkgo	$150	66.8
St Johns Wort	$140	189.7
Ginseng	$ 96	11.2

Garlic	$ 84	17.3
Echinacea	$ 70	41.5
Saw Palmetto	$ 32	74.0
Kava Kava	$ 16	461.5
Pycmpgemp/Grape seed	$ 12	21.5
Cranberry	$ 10	67.7
Valerian Root	$ 8	41.5

Statistics show that in the U.S the majority of herbs are sold in health food stores, but the mass market is catching up very fast, as are mail orders and direct sales that constitute a third of the market. In other parts of the world, Asia for example, the trend is just the opposite. Direct sales are number one, followed by mass market and health food stores. Thus, when one develops a product for global markets, one must remember that distribution channels vary from region to region; one formula does not fit all.

What are the reasons for the slowing growth of herbal sales in recent years? Even though most of these statistics are in the U.S., similar trends appear in other major markets (Europe and in Japan). In the last 10 years consumers have shown tremendous faith and enthusiasm for these types of products and are willing to take them for months without asking for definitive effects. This same type of tolerance is definitely not there for western drugs or OTC products. Consumers may now demand a better performance from herbal products. There is also a public perception of poor quality control that is often driven by price pressures.

But is the public perception contributing to the decline? Recently, media and various interest groups have begun relentlessly to attack the herbal industry making it sound as though the entire industry is lacking quality. As a matter of fact, many companies and associations are doing a great job in this area. In the last couple years, the media has investigated herbal remedies and has drawn negative conclusions about quality, drug interactions, and the side effects of herbal products. Herbal products that show promising results are rarely reported on. Biased reporting may well contribute to the down turn of market.

What are some of the obstacles towards the mainstreaming of herbal medicines?
1. Most of herbal products still heavily rely on anecdotal data.
2. Randomized controlled trials and data are lacking.
3. Many of the products have very vague and broad claims that are especially troubling to the professionals who are not able to understand or grasp what kind of endpoint or health benefit these products may deliver

to their patients, consumers, or clients. That is why clinical study is so important. We need to have very clearly defined end points.

4. A panacea "cure all" phenomenon. Many times we see herbals being talked about which are supposed to have all kinds of health benefits. Some of these maybe true; some partially true; some are exaggerated. A "cure all" claim made over and over again causes loss of confidence in the product.

5. Lack of quality assurance. This is partially due to lack of know how, partially due to price pressure, partially due to opportunists who really don't care about quality, and who just want to make a quick buck.

All this is contributing to the overall image of TCM industry that we hope to change.

2. REGULATORY ENVIRONMENT OF DIETARY SUPPLEMENTS AND HERBAL DRUGS

As stated earlier, TCM can be developed potentially into three types of product. One is as a dietary supplement; the second is as an OTC drug; the third is as a prescription Rx drug.

Table 3. Commercialization of Herbs and Drugs: Regulatory Differences

Dietary Supplements	Drugs
• FDA approval not required	• FDA approval required
• Mixture of active constitutes	• Single well-characterized chemical entities
• Efficacy based mainly on historical and anecdotal data	• Prospective Phase I-III studies
• No GMP guidelines	• Clear GMP guidelines
• Difficult analytical methods	• Well established analytical methods
• Preponderance of evidence	• Unequivocal clinical evidence

Table 3 illustrates the regulatory differences between dietary supplements and drugs. In the U.S., commercialization of dietary supplements does not need FDA approval, whereas marketing of drugs and OTC products does. And this morning we have seen some of the guidelines governing the OTC drugs develop as well as those for prescription drugs. So that is very good news. Getting FDA approval doesn't automatically bring in a market share of a product. There's still a lot of marketing and education needed, and maybe a partnering strategy needs to be taken into consideration.

Dietary supplements usually have a multiple active component, and the Rx/OTC drugs usually have very well defined mechanisms, and clear active components, single and well characterized. The efficacy of dietary supplements is based mainly on historical anecdotal data. For drugs you have to conduct Phase I, II, and III studies, which will cost a lot of money. Dietary supplements usually are limited to oral dosage form. We've seen many companies try to get around that by using nasal or sublingual patches that are actually not legal. Strictly speaking, dietary supplements have to be absorbed through the GI track. But with the drugs: OTC, Rx, you can have all kinds of forms, injections, patches, etc. There is no GMP guideline for dietary supplements, and there is a very clear GMP for drugs. The dietary supplement's analytical methods are very difficult, but modern science and tools now available can do many things to overcome that problem.

3. INTEGRATION OF DIETARY SUPPLEMENTS INTO MEDICINE

For the TCM to develop as drugs or as dietary supplements, what really needs to be done? First let's talk about product development for TCM to get into mainstream health care. There is a cultural expectation and a lack of credibility that is in my view the single most difficult problem in realizing a profitable market. Using herbal products, including TCM, for treating disease is well established in Europe. In Japan, a very large percentage of people are comfortable in using herbal products as part of their health care, but in the U.S. those people are in the minority. A very small percentage of the population in the US is willing to go out and try. They believe that Chinese medicine is genuine; it's real; it's original, and it works. But they also believe that Chinese medicine, especially the patented medicine in formulated form, may be adulterated, may be of poor quality, may have toxic material, and it may be that the efficacy is due to the combination of TCM with some western medicine. That is an image we have to turn it around. We have to make people believe that Chinese medicine has true merit on its own. To develop traditional Chinese medicine for the western market, governments, especially the oriental, (China, Taiwan, Singapore, Hong Kong) should get together to think about a strategy to include their curriculums in medical schools: medicinal chemistry, natural products, the terminology for traditional Chinese medicine. Most professionals would then have some understanding or familiarity or some comfort level in dealing with consumers or patients who come to them and say " I heard about this, I want to try this, what do you think?" TCM would not be completely unknown to them. This is a long road. We really need to do a lot for this to happen. Developing awareness should run parallel to product development. You can spend five or six years developing a product. If no mainstream doctors accept TCM, it won't have a very good market or sales number anyway. The second part is manufacturing. Ideally, a regulatory body should develop some kind of guidelines for GMP as well as standards for traditional Chinese medicine. And such standards should

174

be globally recognized standards. Such standards will help the development of the market and bolster confidence in the product.

We know traditionally it takes 7-10 years to develop a new drug, even with the new FDA guidelines to allow parallel development of Phase I and II clinical trials. It still will take a long time to bring an herbal product to the market as a drug. And thus, it might be wise to market products as dietary

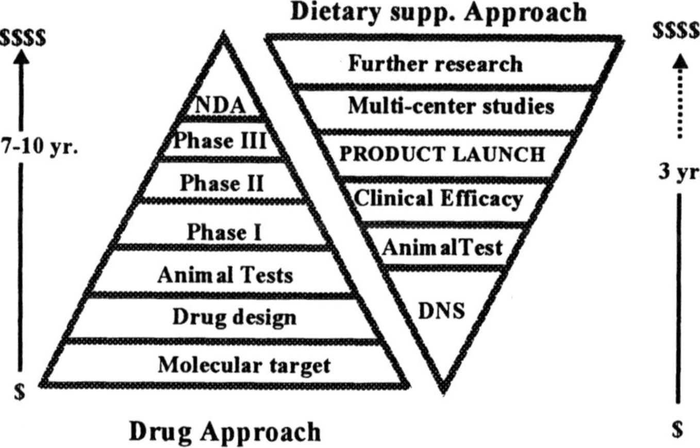

Figure 1. Dietary supplement vs. drug development

supplements. In the mean time, we can develop the product into an OTC or prescription drug (see Fig. 1). It is possible tolaunch the product very rapidly, and then do most of the development after the product is on the market. The amount of money required is very large to develop a new drug, even with a TCM herbal product. By marketing TCM as a dietary supplement, it will save time and money. There is a ten to one difference in terms of these two development schemes. But there are some perils in this strategy. One is that if a product is developed as a dietary supplement, and at a later stage switch to a prescription, there might be a public perception problem. Thus, it is critical to consider all aspects of product development prior to launching a product.

Table 4 Stereotype of Herbal Remedies
- Herbs are natural and therefore safe
- Herbs do not have side effects
- Herbs are a panacea
- Drug interactions do not occur with herbal remedies
- If one herb is good, two herbs are better

Table 4 lists some of the stereotypes about herbal remedies that we need to work very hard to change. First, most people think of herbs as natural, and therefore safe. Nothing could be farther from the truth. We have studied herbal products in the pharmaceutical industry for many years. We know that 70% of all herbal products have some toxic substance, and cause some side effects. Very few of them are truly safe. It all depends on the dosage. It depends on how much you take and for how long. Second, there is a perception that herbals do not have side effects. Again, this is not true. Herbs are chemicals, and they are made by nature. And since they are interacting with physiological systems on many sites, they've got to have some impact. It could be a negative side effect; it could be mild and tolerable. Herbal medicine is often thought of as a panacea in that it can do a lot of things, treat many diseases. Again, that may be true, or it may not be. It all depends on the dosage, the duration, and the clinical symptoms. The philosophy of TCM, being multiple herbs, is that they tend to interact with multiple targets in your body. Therefore, they may in fact have more than one effect. But it all depends again on the dosage, duration and usage. In terms of drug interactions, herbs interact with each other and with other drugs. It's wrong also to believe that if one herb is good, then two herbs are better. When we develop a product, we really need to be very much endpoint oriented, and focused on what a product really can do to help specific conditions.

Table 5. Clinical research and dietary supplement products

Research objective	Level of difficulty
• Safety	Low
• General population	Moderate
• Mechanism of action	Moderate to high
• Pharmacokinetic	high
• Clinical pharmacology	High
• Efficacy	Moderate to high

Table 5 compares different degrees of difficulties in clinical research on herbal products in terms of various aspects of product development for commercialization. On the left side, 6 areas need to be considered. The most important one is safety. We have to prove that the product can be used safely, if it is used according to the instructions. That is easy to do. Traditionally, TCM doctors develop a specialized formulation for each patient. Is it possible to develop a product for the general population: male and female, young and old, heavy and skinny (250 pounds to less than 100)? Yes, it would not be too difficult. By changing the formulation, multiplying it a little bit, I think you can achieve a reasonable dosage. In order to gain a mass market this has to happen. Otherwise, we can offer only highly specialized products. Third is the mechanism of action. That is difficult. But with modern tools such as gene chips and an understanding of molecular biology, that can be achieved, or partially achieved anyway. So mechanism of action is achievable these days. And difficulty is moderate to high. Pharmaco-kinetics, that is understanding the drug levels metabolized, and the tissue distribution is very difficult. And that probably will remain difficult for a while. At this point, the FDA is willing to overlook this part and work with the industry to try to develop a product first. Clinical pharmacology is also very difficult. This last category is the issue of efficacy that relates to the clinical endpoint, the outcome. How do you design a clinical trial to reflect the true benefit of TCM in western terms? TCM can be developed into drugs or even dietary supplements. There are different levels of difficulty that make product development very challenging, yet it's a fun thing to do. And if you can do well, you can generate a good business.

4. CONCLUSION

The key to developing TCM products is to think globally. There are tremendous opportunities for TCM in the long run. There should be a place in the mainstream health care for TCM in the next century. It will take time but it is possible to make progress by tackling some of the simple safety and quality issues so that it can occupy a portion of the market in dietary supplement, functional food, etc. And with effort, working with the FDA and other regulatory bodies, eventually, TCM can break into OTC and prescription drugs in a modest way. Finally, to realize the large global market, there are a few things that are musts. First is the lobbying and education to gain credibility of professionals in the developed market such as Europe, Japan, and US. Second, the manufactures of herbal products must work with the regulatory agents to finalize a good system, a reasonable system to build this product. Even though the consumer price pressure is very high for global products, companies cannot charge a lot of money for these products. Finally, there must be a good link between east and west to understand the health benefits of TCM. How can we translate eastern philosophies such as multiple target points, the balance of nature, yin and yang into acceptable terms for the western culture? Once the consumer truly feels comfortable many of these products will be accepted and utilized broadly. So, it is still hopeful that TCM will develop and will occupy a great portion of the future market. But there are obstacles that will require the investment of work, time, energy, and money before that can happen. So be ready. If one wishes to play this game, be patient, and also develop a good strategy.

Chapter 19

TRADITIONAL CHINESE MEDICINES:
REGULATORY AND SCIENTIFIC CHALLENGES

MARK EDGAR
Chemistry, Manufacturing and Control, Ancile Pharmaceuticals, Inc.
San Diego, CA

Abstract:
Development of multi-component botanical products as pharmaceutical drugs requires control of consistency. The foundation for consistency, which provides the basis for establishing comprehensive specifications, is based on control of the biomass and thorough characterization of the extract. Validation of the manufacturing process ensures that the specifications are reproducible. Traditional Chinese Medicines require a special focus on defining the composition of the pharmaceutical product to be developed, understanding the biomass pretreatment chemistry, characterizing the multiple herb mixture, and conforming to drug Good Manufacturing Practices (GMP).

Keywords:
Botanical, herbal medicine, GMP, Traditional Chinese Medicine, pharmaceutical drugs, validation, fingerprinting, and Chemistry, Manufacturing and Control.

1. INTRODUCTION

Medicines derived from plant sources have been used for centuries to treat various medical conditions. Herbal medicines are often comprised of several different chemical constituents that can interact with different pharmacological targets to afford a multifaceted mechanism of action. This is in contrast to more traditional pharmaceutical drugs, which focus on a single new chemical entity (NCE) and a more well-defined mechanism of action. Bioavailability may benefit from the multi-component nature of botanical products through a synergistic effect between constituents that results in therapeutic advantages when compared to individual isolated compounds. In

Yuan Lin (ed.), Drug Discovery and Traditional Chinese
Medicine: Science, Regulatory and Globalization, 177-190.
©2001 *Kluwer Academic Publishers. Printed in the Netherlands.*

particular, diseases with a multi-factorial cause may benefit from herbal medicines. The therapeutic benefit and safety of many herbal medicines have been qualitatively established, but their potential as new drugs have not been fully realized due to the lack of economic incentives, absence of strict manufacturing controls, and lack of rigorous proof of efficacy. Prior to 1997 there was no regulatory pathway for approval of botanical drugs in the United States. Today, a pathway for drug approval and market exclusivity does exist. In August 2000, the FDA issued Draft Guidance for Industry on Botanical Drug Products, which outlines the pathway to regulatory approval of botanical drugs using the new drug application (NDA) process within the FDA. This further confirms that the regulatory pathway for botanical drugs will indeed be accepted within the regulatory framework of the United States.

The purpose of this paper is to outline some of the challenges associated with developing botanical medicines as pharmaceutical drugs. First, there is a comparison of the botanical drug development to the pathway for NCEs, followed by a summary of the FDA requirements for development of botanical drugs recently outlined in the Draft Botanical Drug Product Guidance. Then, there is a discussion of the chemistry, manufacturing and controls (CMC) challenges represented in those FDA requirements and an overview of Ancile's approach to developing the CMC aspects of botanical drugs. Lastly, the paper outlines some additional challenges for Traditional Chinese Medicine (TCM) development.

Ancile Pharmaceuticals, Inc., founded in 1998, is an ethical pharmaceutical company that seeks to become the leader in development and commercialization of proprietary botanical drugs for the prevention and treatment of disease. Ancile's staff of 45 employees includes 35 physicians and scientists. Laboratories are housed in San Diego, California in the US and in Shanghai, China. Ancile selects botanical products for development based on the strength of clinical data in the literature, the marketing opportunity, the potential to establish a proprietary position. and the potential to produce a consistent product. Pharmaceutical improvements are introduced to the herbal medicine selected for development, including safety and clinical efficacy demonstration, botanical product standardization, Good Manufacturing Practices (GMP), advanced formulation development, and establishment of Intellectual Property. Ancile currently has three botanical products in development in the US, each under an allowed investigational new drug application.

2. BOTANICAL DRUG AND NEW CHEMICAL ENTITIES

Comparison of a botanical drug with NCE drugs illustrates that the main difference lies in the CMC area (Figure 1). The clinical program for a botanical drug and an NCE is basically the same with one exception and is driven by the indication. The clinical program for a botanical drug can begin

	Botanical Drug vs. NCE
Clinical	Program may begin without toxicology if doses are at traditional level
Non-clinical toxicology	Toxicology delayed until Phase 3 if doses are at traditional level
Biopharmaceutics	Requires justification of component(s) used to assess PK
Chemistry, Manufacturing and Control	Requires new approach to justify consistency, strength, purity, identity

Figure 1. Botanical drug and NCE comparison.

at Phase 2, without the usual toxicology program if the doses are at the traditional levels used for the herbal medicine. The non-clinical toxicology program required for a botanical drug will be equivalent to an NCE; however, the timing of certain studies can be delayed and need not start until the Phase 3 clinical program begins. Again, this is true if the doses to be used in the clinic are equivalent to those traditionally used. The pharmacokinetic studies for a botanical product require justification of the constituents to be used. These constituents can include active or inactive components found in the botanical.

The CMC regulatory requirements for a botanical NDA were recently outlined in the Draft Botanical Drug Product Guidance issued by the FDA and are basically equivalent to those prescribed for an NCE with the exception of control of the biomass, which is the source for a botanical product. The overall goal for the requirements is to justify consistency, strength, purity, and identity for the drug substance and drug product. While the CMC regulatory requirements for botanical drugs and NCEs are essentially the same, demonstration of consistency, strength, purity, and identity for a botanical product requires a new approach because of the multi-component nature of these products. NDA approval of a botanical product requires a validated manufacturing process, in-process controls, demonstrated lot-to-lot consistency, specifications, validated analytical methods, validated stability-indicating methods, and conformance to GMPs.

One of the greatest challenges for a botanical drug is demonstration of lot-to-lot consistency. Botanical raw material source (biomass) is inherently variable depending on the cultivation conditions that were present during the growing cycle. For instance, drought, excess rain, temperature conditions and other environmental conditions can have an influence on the chemical composition of the biomass, which, in turn, influences the consistency the botanical drug substance. Demonstration of lot-to-lot consistency is very important for NDA approval. The foundation for demonstrating lot-to-lot consistency is the specifications that are established during the development process. These specifications define key attributes for the botanical product and must include acceptance criteria. For example, if an active constituent has been defined for a botanical product, then a concentration range must be established for that active constituent in the botanical drug product. Each lot of the botanical material produced must meet that specific range in order to be acceptable for use.

The parameters defined as part of the specifications require analytical methods to assess whether a particular botanical product lot meets those specifications. These analytical methods must be validated according to FDA guidelines. As previously mentioned, the specifications will include an assessment of active and inactive constituents in the botanical product. The analytical methods used to measure the level of active or inactive constituents within a botanical product require a reference standard as part of the analytical process. These reference standards are pure samples of the individual constituents. For botanical drugs, several reference standards are often required. This can be particularly challenging for a botanical product because the reference standards often must be isolated from the botanical material itself. In addition to the pure compounds required as reference standards, reference standards will be required for the biomass used to produce the extract, the botanical drug substance, and the drug product.

An important consideration in developing any drug for NDA approval is the demonstration that the drug is stable over the shelf life of the product. This requires development of stability-indicating analytical methods. These methods must be validated as well.

The final requirement for NDA approval involves production of the botanical drug substance and botanical product. All production must be conducted using GMPs. The FDA will inspect the facilities and procedures used to produce the botanical product prior to NDA approval.

3. CMC CHALLENGES

To summarize, the CMC challenges for NDA approval include controlling the biomass and assuring consistency of a complex mixture. Ensuring consistency involves characterizing the extract, establishing specifications to reflect that characterization, and establishing that the process is reproducible through process validation. The stability of the drug substance and drug product must be demonstrated for NDA approval, and this is particularly challenging due to the multi-component nature of a botanical material. Issuance of the botanical guidelines has certainly helped define a pathway for regulatory approval, but since no NDA approval has been granted for a botanical drug, this pathway is unproven at present.

4. ANCILE'S APPROACH

Ancile's approach to botanical drug development includes adherence to Good Agricultural Practices (GAP). GAP involves understanding the growing conditions required to yield a reproducible biomass, which, in turn, serves as the source for the botanical drug substance and, ultimately, the marketed drug product. A typical GAP program documents, for example, the qualifications of the grower used to produce the biomass, the source of the seeds, seed storage, the cultivation process, and the harvest and preparation of the biomass for extraction. Documented activities may also include raising seedlings in a greenhouse, transplanting the seedlings into the fields, and developing biomass cleaning and drying procedures. The instructions for biomass production are prepared prior to beginning cultivation. The actual procedures used during the cultivation process are documented as well. As part of the qualification process for a particular grower, the history of land use is documented and includes information concerning crops previously planted in the fields and the chemicals used in the prior three years. These chemicals might include various pesticides or herbicides. Lastly, specifications with acceptance criteria must be established for the biomass. These specifications might include pesticide levels, heavy metal contamination, presence of active constituents, characteristic markers associated with the particular plant, and microbial levels.

5. BOTANICAL DRUG CHARACTERIZATION

Once the biomass is available, the drug substance is typically produced through an extraction process. The extract generated must be well characterized in terms of its biological activity and chemical content. This

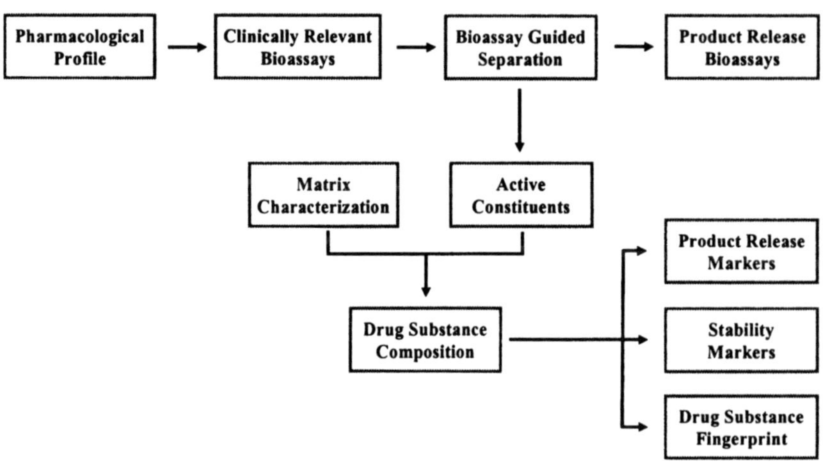

Figure 2. Characterization flowchart.

characterization process begins (Figure 2) with a broad pharmacological profile of the extract, which is compared to the possible mechanism(s) of action for the clinical indication to be developed. This procedure defines the bioassays that are relevant to the clinical program and are subsequently used to identify the active constituents. The procedure involves separating the extract into its component chemical constituents using a variety of separation techniques, including normal phase chromatography, reverse phase chromatography, and solvent partitioning. The clinically relevant bioassays are used to determine the bioactivity of the fractions produced by the separation process. The separation process continues until pure active chemical constituents are identified. Occasionally, no individual constituents can be identified as active components. In this case, other inactive (matrix) constituents can be used as markers for specifications. Some constituents also may serve to enhance bioavailability but are not active with respect to the clinically relevant bioassays. These matrix constituents are identified and quantified where possible. The combination of the active constituents and the matrix components define the drug substance composition. The goal for this characterization process is to identify as many of the extract constituents as possible. The composition of the drug substance serves as the foundation for a number of regulatory activities that include defining markers for product release, stability, and the pharmacokinetic program. The overall composition of the drug substance can serve as a fingerprint, a regulatory requirement defined in the Botanical Drug Product Guidance. The clinically relevant bioassays used in the bioassay-guided separation of the extract also can serve as product release bioassays, which is a suggested specification in the Botanical Drug Product Guidance. The extract specifications will control composition, bioactivity, and possible contaminates. The active and inactive constituents define the composition. The specification for bioactivity will be determined by a bioassay that is relevant to the clinical indication. Specifications for possible contaminants include pesticides, heavy metals, microbial contamination, and solvents that are used in the extraction process.

The FDA has devised analytical method (FDA Pesticide Analytical Manual, Volume I, 3rd Edition, Section 302) capable of measuring up to 200 widely used pesticide residues, including some pesticides that are only used outside the USA. Chemical classes measured including carbamates, organochlorides, organonitrates, organophosphates, organsosulfer compounds, and pyrethroids.

6. CHEMICAL FINGERPRINTING

Establishment of a chemical fingerprint, as required in the Botanical Drug Product Guidance, represents a significant challenge. The extract of a botanical material is usually composed of many different kinds of compounds. These compounds might include fatty acids, amino acids, terpenes, very polar materials, and nonpolar materials. Each class of components may require a different type of analytical method to assess their content. A fingerprint, as

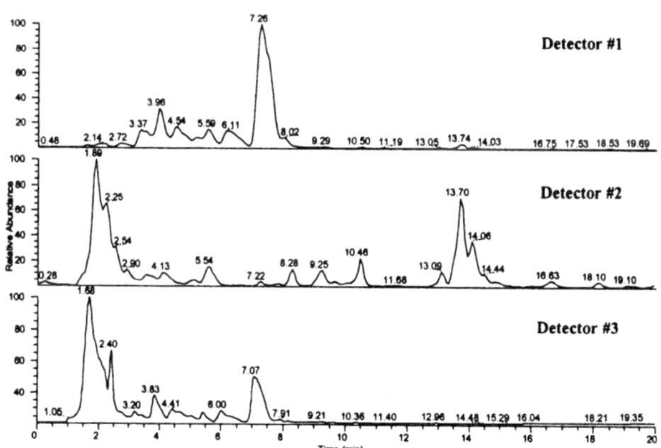

Figure 3. Fingerprinting example: Ancile's Multi-Dimensional Profiling™

defined by the Botanical Drug Product Guidance, may include various HPLC separations identified by different detectors. The sum of these separate analyses make up the fingerprint of a botanical extract. The simple example shown in Figure 3 includes only a single HPLC separation analyzed using three different detectors. As can be seen, each detector yields significantly different information. The total fingerprint could be considered to be the sum of the three scans. The challenge for chemical fingerprinting lies in developing a methodology that will allow meaningful specifications to be established. A meaningful specification requires quantitative comparison of one extract fingerprint to the fingerprint of another extract. To execute this comparison, a specific extract must be established as a reference standard, and all subsequent lots of botanical material compared to that reference standard. A typical chemical fingerprint might be composed of a hundred or so separate HPLC peaks, and each peak would have an inherent variability associated with it. Establishing a specification involves defining an allowable range of variation for peak-to-peak comparisons then consolidating all the individual comparison data into an overall specification that allows a definable pass/fail limit to be set. The Botanical Drug Product Guidance also requires a qualitative assessment for fingerprint comparisons. Again, given the complexity of the relatively simple chemical fingerprint, shown in this example, it would be very difficult to say qualitatively that one chemical fingerprint is equivalent to another chemical fingerprint. To be useful, we believe that a quantitative fingerprint is required and comparison methodology must be established to systematically deal with the complex data. Ancile's Multi-Dimensional Profiling™ methodology facilitates these complex comparisons.

7. PROCESS VALIDATION

Process validation is an important aspect of drug development and is particularly important in the area of botanical drugs. Minimizing the variability in the process used to produce an extract and a drug product is very important. Essentially, process validation is defining what parameters are important to maintain consistency of the manufacturing process and establishing the necessary controls to ensure consistency. Identification of those critical parameters is based on experience and scientific judgment. Once the critical parameters have been identified, the limits for those parameters are established or justified in a laboratory setting. The goal for process justification is to establish the outer boundaries for critical parameters that will yield acceptable product. Validation of the manufacturing process is the demonstration of reproducibility within the confines of the defined process parameters.

For example, laboratory experimentation may show that extraction of biomass functions satisfactorily between the temperature ranges of 50° and 80° (Figure 4). The temperatures for the extraction specified in the New Drug

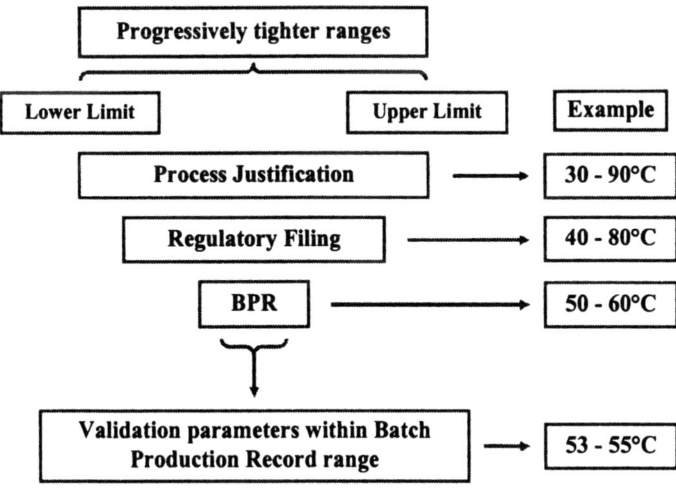

Figure 4. Process validation methodology.

Application filing with the FDA might specify a narrower range to ensure that the process is always within the experimentally proven range that will provide a reproducible extract. The batch production record (BPR) used in routine manufacturing would have an even tighter range for the extraction process temperature to ensure that the extraction process operates well within the range specified in the regulatory filing and again, well within the range justified as part of the laboratory experimentation. Demonstration of consistency or process validation would occur within the temperature ranges specified by the batch record. Specifications are used to assess success of the validation effort.

8. STABILITY

Stability of the drug product is an important aspect in any drug development program. The stability program typically uses ICH guidelines as the foundation for the stability parameters and methodologies used.

Development of stability-indicating assays is critical to the success of the program. Parameters that can be monitored as part of the stability program include active constituents, key inactive constituents, specific degradants, microbial specifications, moisture, and appearance.

9. TRADITIONAL CHINESE MEDICINES

Traditional Chinese Medicines are often composed of multiple herbs, which have been selected by a Traditional Chinese Medicine (TCM) practitioner to treat specific diseases in individual patients. The medicines are tailored for individual patients and the botanical components of those medicines can change accordingly. Since TCMs are "individualized", converting a TCM to a pharmaceutical product is particularly challenging because a pharmaceutical drug requires a specific composition. Defining a specific composition can be tricky because the efficacy of a TCM preparation may be used over a range of compositions. Overall, developing TCM as pharmaceutical drugs presents unique challenges, including coping with multiple herbs, controlling the biomass, which often comes from China, maintaining control over biomass pre-processing, which includes steps uniquely associated with TCMs, converting a TCM to a pharmaceutical product, and adhering to Good Manufacturing Practices during the production process.

10. TCM PREPARATION

Preparation of a TCM begins with the cultivation of the biomass, harvest of that biomass, and pre-preparation of the biomass using unique processing parameters in certain cases. The remainder of the process – the extraction and formulation – is usually undertaken by the patient. The patient receives the prescription from the TCM practitioner, which can include a number of different herbs. These herbs are prepared by the patient, usually through a water extraction (making tea), and then consumed.

Representatives from Ancile Pharmaceuticals visited a number of TCM farms in China and found these farms to be extremely well run and controlled. A typical farm involves cultivation of a number of different types of agricultural products, including rice and herbs for botanical products. Once harvested, the biomass is transported to a central processing facility and pretreated in preparation for the extraction process.

Biomass pretreatment can include the following: heating for two or three days, smoldering until black, heating with wine until dark, and heating with moxibusted bran until a certain biomass color is achieved. One pretreatment process prescribes boiling the biomass with vinegar for several hours and then drying. This process may be associated with increasing alkaloids solubility prior to extracting the biomass. The chemical modifications that may occur during these pretreatment steps could be significant. Heating biomass at high temperatures could cause

Figure 5. Herb pretreatment facility in China.

chemical decomposition of the constituents; boiling with wine or vinegar certainly would cause some chemical changes. If a biomass pretreatment process is used to manufacture a pharmaceutical drug, the process must be controlled to ensure reproducibility. Ancile has inspected biomass pretreatment facilities in China, and a picture of one facility is shown in Figure 5. In the foreground is a dryer, which is like a coffee roaster where biomass is heated. In the background is a large wok that is used to process the biomass with other chemical constituents, such as vinegar or moxibusted bran. To establish controls for biomass pretreatment, the nature of the chemical changes caused by this step need to be determined.

Once the chemistry is understood, then controls can be established and reproducibility demonstrated. For example, process validation could be used to establish reproducibility for pre-processing. Fingerprinting may be useful given the complex nature of the chemical changes that are likely to occur. The facilities used for biomass pretreatment are multi-purpose and process many different kinds of herbs. It is important that procedures be in place to control potential cross-contamination between various herbs. Cleaning procedures and documentation become very important.

11. PHARMACEUTICAL TCM FORMULATIONS

The formulation used for TCM is typically a tea produced by the patient. The herbs are covered with water, allowed to soak at room temperature for several minutes, heated to boil and filtered. The resulting tea is the drug product. Issues associated with a traditional tea include regulatory acceptability, Western preferences, ease and convenience of administration, and formulation efficacy. Typically, for a pharmaceutical drug, the drug product is defined, for example, as a pill or a preformulated liquid that the patient simply takes. In the case of a TCM, the patient is actually generating the drug product through an extraction process. From a regulatory standpoint, this process must be reproducible so the drug product that results is the same each and every time. Because of this, having the patient conduct a direct extraction of the biomass may not be feasible.

Western preferences for a pharmaceutical dosage form do not usually include a tea. The most preferred formulation is a solid oral dosage. Converting a TCM tea to a solid oral dosage form can potentially affect the efficacy of the TCM. An aqueous extract of biomass can produce a fairly large extract mass because the extract includes large quantities of inert

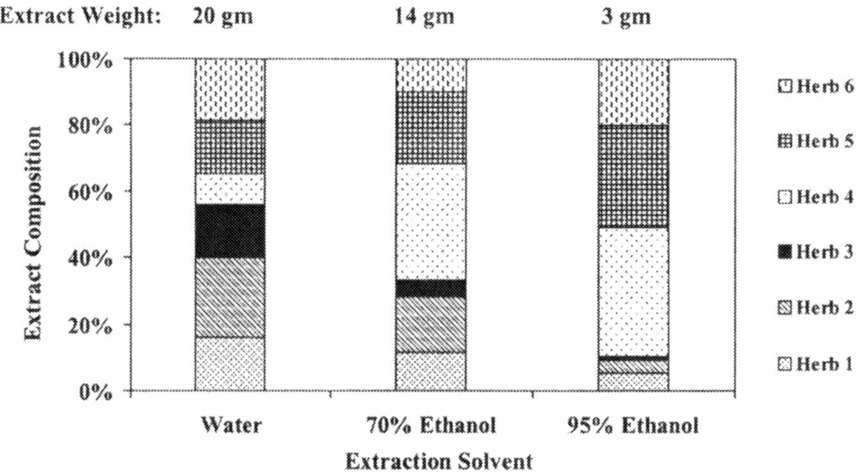

Figure 6. Impact of extraction methodology on TCM composition.

materials. Minimizing the extract volume will change the chemical content of the traditional tea, which in turn may impact efficacy. An example of the impact of minimizing the extract volume is shown in Figure 6. Here, a water extract of a six-herb mixture yields an extract weight of about 20g and has a specific composition for the six herbs. If the solvent is changed from pure water to 70% ethanol and water, the volume of the extract goes from 20g to 14g, but the composition of the extract changes dramatically as some herbs make up a much greater proportion of the extract, while others diminish. If the extract solvent is changed from 70% ethanol to 95% ethanol, the extract volume declines from 14g to 3g, and the extract composition also changes greatly. In this case, three of the six herbs become dominant, where the other three herbs are greatly diminished. As a result, the efficacy of the 3g extract compared to the 20g extract could be significantly different.

12. TCM AND GOOD MANUFACTURING PRACTICES

In order for a TCM to be manufactured in China and distributed in the United States, all the manufacturing facilities utilized in the production process in China would be inspected by the FDA for conformance to GMP used for pharmaceutical manufacturing. One of the key GMP activities is documentation of the manufacturing process throughout the manufacturing chain. For production in China, several issues can occur, such as language translation and structure and format of the documents used in the manufacturing process. There are specific requirements for GMP documentation that must be met. Those requirements must be translated into documents that can be used in China. As part of the documentation process, the use of pesticides needs to be defined and controlled. Translation issues can make determining the identity of pesticides difficult. Documentation of the fertilization practices would be required as part of the GMP process.

For example, review of certain Chinese documentation shows that human waste can be used as a fertilizer for some herbal products. This practice would not be acceptable for a pharmaceutical drug regulated by the FDA. Documentation of the fertilization practices would be required as part of the GMP process.

13. CONCLUSION

In conclusion, the foundation for herbal product consistency is based on control of the biomass and thorough characterization of the extract. This foundation provides the basis for establishing comprehensive specifications. Validation of the manufacturing process ensures that the specifications can be reproducibly met. TCMs require a special focus on defining the pharmaceutical product that will be developed, understanding the biomass pretreatment chemistry, characterizing the multiple herb mixture to form a good foundation for specifications, and conforming to drug GMPs.

ACKNOWLEDGEMENTS

I would like to acknowledge the contributions of the entire Ancile Team and especially Mr. Michael Andrews, Dr. Amaresh Basu, Dr. Brad Carte, Ms. Karen Church, Dr. Howard Constant, and Dr. Janice Thompson.

Chapter 20

MARCO POLO TECHNOLOGIES:
A MODEL FOR A MODERN TCM COMPANY

ROBERT YUAN
Department of Cell Biology and Molecular Genetics
University of Maryland, College Park, MD 20742

Abstract: A major challenge today is the integration of Traditional Chinese Medicines (TCM), particularly herbal preparations, into the global pharmaceutical industry. This paper explains the approach used by a modern TCM Company, Marco Polo Technologies (MPT), working with partners and collaborators in the US, China, and Taiwan. MPT's ultimate objective is to develop FDA approved therapeutics for the treatment of chronic diseases. The company's strategy involves the early development of standardized TCM that can be marketed as nutritional supplements in both the US and Asia. The development of FDA therapeutics is based on combinatorial medicines that have a mixture of active compounds. MPT's activities, including acquiring raw materials, product development, manufacturing, and clinical trials, involve a significant amount of work in China and Taiwan. MPT has now developed three products for skin infections, allergies, and arthritis. In order to gain approval for these products, MPT had to overcome numerous hurdles presented by the FDA and the Patent Office.

1. OBJECTIVES

The growing cost and complexity of modern Western drugs has led to a search for alternative approaches. In the US and Western Europe there has been the rapid growth of herbal medicines. The WHO has launched a new initiative in the use of traditional medical remedies, and the Chinese government has established a new program to promote the modernization of TCM and their use.

Marco Polo Technologies (MPT) is a startup biotechnology company that is dedicated to the establishment of a number of product platforms for health

Yuan Lin (ed.), Drug Discovery and Traditional Chinese
Medicine: Science, Regulatory and Globalization, 191-200.
©2001 *Kluwer Academic Publishers. Printed in the Netherlands.*

care products. Such product platforms are used to develop a broad spectrum of healthcare products (for both U.S. and Asian markets) ultimately leading to FDA approved therapeutics. MPT differs from other companies that have similar objectives in the following ways:

- It operates as a virtual company with close collaboration with US and Asian research laboratories and companies.

- It focuses on chronic or recurring diseases.

- Its therapeutics is based on combinations of active compounds rather than a single chemical entity.

2. MPT's THREE-PRONGED R&D STATEGY

2.1 DEVELOPMENT OF STANDARDIZED TCM PRODUCTS

The first part of the strategy involves the development of standardized Traditional Chinese Medicine products (sTCM). The careful staging of a selection of TCM formulations, chemical standardization of major components and biological assays is leading to the short-term launch of sTCM as nutritional supplements. Such products have rapidly growing markets both in the U.S. and Asia. The Dietary Supplement Health and Education Act of 1994 allows botanicals to be sold in the U.S. as nutritional supplements, which are not subject to FDA drug regulations. Standardized TCM can be sold taking advantage of this law as long as no claims to be a drug are made on the label. Open label clinical studies are being used as a powerful tool to differentiate MPT's products from those of its competitors.

2.2 A NOVEL DRUG DISCOVERY PARADIGM BASED ON COMBICINES

The second part of the strategy is a novel drug discovery paradigm based on combicines, combinations of active compounds derived from. MPT is patenting such combicines based on their chemical standardization, biological assays, and novel use for the treatment of specific diseases. MPT is undertaking the R&D and early stages of clinical studies, while the multi-center pivotal clinical trials for the FDA approved therapeutics based on such combicines will be done through strategic alliances with major biotechnology and pharmaceutical companies. In the long-term, MPT hopes to develop, manufacture and market some of its own combicines.

2.3 DEVELOPMENT OF NON-PHARMACEUTICAL APPLICATIONS OF COMBICINES

The third prong of the strategy is the development of non-pharmaceutical applications of combicines. For example, one of MPT's TCM formulations is used in the preparation of products for skin care. It also lends itself to veterinary and agricultural applications. The licensing of non-pharmaceutical applications takes advantage of such opportunities to generate income without significant diversion of resources.

3. BUSINESS STRATEGIES

Given the nature of TCM, MPT will focus on the research, product development, clinical testing, regulatory approval, and patent protection of its products. In collaboration with corporate partners in the US and Asia, MPT will initially manufacture and market nutritional supplements. By generating revenues and a deeper understanding of the mechanisms of action of such TCM products, MPT will work with larger corporations to take such products through drug approval.

A fundamental basis of MPT is it will be a transPacific business that will be flexible in maximizing both the resources and markets in the US and Asia. MPT has implemented its strategy in the following manner:

3.1 OBTAINING RAW MATERIALS FOR MPT'S PRODUCTS

The herbs that are the raw materials for MPT's products show considerable variations from lot to lot. Therefore MPT has established a network of Asian and US suppliers that can provide both raw herbs and extracts that meet MPT's specifications. Multiple suppliers are essential in being able to obtain lots of raw material with the right content of active compounds.

3.2 RESEARCH: CHINA AND THE US

The formulations best suited for a particular clinical condition are identified and chemical analyses done in Chinese laboratories. Then, in the US, the herbal extracts formulated by MPT are subjected to chemical standardization and biological assays. The characteristics of the individual actives as well as those of the combicines are determined. Research into the mechanisms of action of said combicines is currently under way. Specific products and applications are then patented.

3.3 PRODUCT DEVELOPMENTS AND MANUFACTURING

Once a TCM formulation has been chemically and biologically standardized in the US, it is formulated into actual products. Then, pilot lots of active extract are manufactured in China and finished in the U.S. These are the products used for clinical studies and marketing trials.

3.4 CLINICAL TRIALS

Clinical outcome studies are being carried out in China and will also be performed in the US. The data for such studies are being used for the later development of FDA clinical trials as well as for the marketing of the sTCM products, described below. An initial set of clinical trials reporting positive results for MPT's Allerdyne product for the treatment of allergy has been completed in Shanghai. A second set of clinical trials with the topical Rubriderm for skin diseases gave excellent results. A third trial with preliminary positive results for MPT's Arthrene product for the treatment of arthritis has also been completed.

3.5 MARKETING AND SALES

MPT has begun to sell nutritional supplements through agreements with pharmaceutical marketing groups. The sTCM can take advantage of the rapidly growing market for green pharmaceuticals in the United States (over $3 billion in 1997) and generate significant revenues. MPT will recruit a marketing team that will plan and implement different approaches for the introduction of its sTCM into the US market. The licensing of spin-offs, explained in more detail in the Cosmeceuticals section, is an opportunistic effort to generate revenues without any significant diversion of resources. In Asia, MPT is currently in negotiations with companies in Shanghai and Taipei to market sTCM as nutritional supplements in those markets.

3.6 STRATEGIC ALLIANCES

In order to allow for long-term development of its drug products, MPT is pursuing a transPacific strategy to maximize revenues from the sales of nutritional supplements. In the US, MPT is launching the nutritional supplements described above and is also developing a growing portfolio of chemical, biological and clinical data on its products which it intends to patent. Chemical and manufacturing control data, preliminary clinical data, and patents will be used to develop strategic alliances with major pharmaceutical companies interested in converting combicines into FDA approved prescription drugs. It is MPT's intention to complete phase 2 trials on its products in order to maximize their value prior to entering into strategic agreements with major pharmaceutical firms.

In Asia, MPT is in negotiations with pharmaceutical companies in Shanghai and Taipei. The agreements under discussion would have MPT provide products and technology while the Asian partners would source the raw herbs, manufacture and market the products. The objective is to introduce sTCM as US style nutritional supplements and then go through the Chinese drug approval process to have them licensed as Chinese drugs. The 1999 Chinese market for nutritional supplements is growing rapidly, estimated at $2.4-3.6 billion for 1999 and projected to grow to $6.1 billion in 2000.

3.7 REGULATORY APPROVAL AND PATENTS

MPT has considerable expertise on the FDA regulation of complex therapeutic products and of nutritional supplements. MPT has received its first patent for the naturally occurring combination of six compounds that is effective against antibiotic resistant pathogens.

4. sTCM PRODUCT PIPELINE

MPT has established a pipeline of sTCM products, the first generation of which has gone through product development and pilot production (Table 1).

Table 1: Indication and status of six MPT sTCM products

Product	Indication	Status
Allerdyne	Oral, hayfever, mile asthma	Ready for commercial production
Dermagen	Topical, dermatitis	Ready for commercial production, clinical tests
Arthrene	Oral, arthritis	Ready for commercial production, clinical tests
Cholest	Oral, hypercholesteremia	2nd generation- initial R&D

While there are four sTCM products in the first generation, the ones that will be launched into the US, Chinese, and Taiwanese markets will depend on current discussions with corporate partners. MPT and its Chinese collaborators are evaluating formulations for the treatment of late onset diabetes and hypertension. It should be noted that not all sTCM will be selected for further development as combicines. The decision to develop a particular formulation as a FDA prescription drug will be based on the cost of the clinical trials versus the potential market size and value.

5. COMBICINES - A DRUG DISCOVERY PLATFORM

One of MPT's first products, Dermagen, is an ointment used for the treatment of skin infections and atopic dermatitis. This TCM formulation had been reported to have a number of different biological activities. In our laboratories, the extract Rubricine has been shown to have powerful anti-bacterial, anti-fungal, and anti-inflammatory activities. In addition, it also stimulates wound healing, is non-toxic, and is effective against certain antibiotic resistant pathogens. The six active compounds in Rubricine have been shown to act in a synergistic manner.

Rubricine is a combicine product platform that can be formulated into a variety of products that are directed towards different diseases. Some of these diseases are difficult to treat and may be caused by antibiotic-resistant pathogens. MPT has received its first US patent for Rubricine, "Composition for treatment of antibiotic-resistant bacterial infection and method for using and preparing the same." There are numerous formulations for the Rubricine combicine platform (Table 2).

Table 2: The formulations for the Rubricine combicine platform.

Disease	Major Pathogen	Formulation
TOPICAL:		
Otitis media	Streptococcus pneumoniae	Ear drops
Candidiasis	*Candida albicans*	Suppository/cream
Diabetic necrotic sores	Multiple	Ointment
Grey toenail	*Tricophyton ruben*	Nail polish
Athlete's foot	*Tricophyton ruben*	Ointment
Dandruff	*Pitysporum ovale*	Shampoo
Acne	*Proprionibacterium acne*	Ointment
ORAL		
Pneumonia	*Strep. Pneumoniae*	Oral
TB	*Microbacterium. Tuberculosis*	Oral

6. COSMECEUTICALS PIPELINE

One of MPT's first products, Dermagen, is an ointment used for the treatment of skin infections and atopic dermatitis. This TCM formulation had been reported to have a number of different biological activities. In our laboratories, the extract Rubricine has been shown to have powerful anti-bacterial, anti-fungal, and anti-inflammatory activities. Several marketing specialists have proposed that cosmeceuticals based on Rubricine could have wide acceptance as natural and effective products for skin health maintenance. These products have now been formulated and gone through pilot production. The safe, standardized, and wholly organic products that have been developed by MPT for specific dermatological disorders and as cosmeceuticals are each described below.

6.1 RUBRI-DERM

This ***Rubricine*™**-containing ointment is a modern form of a formulation used in China for centuries to soothe and heal skin infections and allergic reactions. ***Rubri-Derm*** adds cocoa butter and vitamin E to further enhance its regenerative power long recognized in the root of the Chinese herb Zi cao.

6.2 FUNGIDEX FOR ATHLETE'S FOOT

Athlete's foot is primarily caused by fungi in the family of *trichophyton*. ***FungiDex*'**, provided in ointment form, exploits the anti-microbial powers of Rubricine™. It is gentle and effective and does not have the severe side effects of some of the drugs often used for this irritating and often lingering problem.

6.3 FUNGIDEX FOR TOENAIL FUNGUS ("GRAY NAILS")

"Gray" toenails are primarily caused by a family of fungi, called *trichophyton*. Provided in the form of a nail polish and exploiting the anti-microbial powers of ***Rubricine*™, *FungiDex*** — when applied regularly — penetrates beneath the surface of the nail, attacking the fungus without the severe side effects of some of the drugs often used to address this vexing disorder.

6.4 DANDROFF™ SHAMPOO AND HAIR TONIC

Dandruff — the scaling of the skin of the scalp — is promoted by a form of yeast, *Pityrosporum ovale*. When applied regularly as shampoo and/or hair tonic, ***DandrOff's*** primary active agent, ***Rubricine*™** kills this yeast, stopping dandruff's unsightly scaling as well as its notorious itching.

7. SPIN-OFFS WITH POTENTIAL MARKET VALUE

Rubricine also has a number of useful spin-offs with large potential market values. Some of these spin-offs include anti-bacterial use in band-aids and the treatment of wounds and skin infections in cats and dogs. Other possibilities include anti-bacterial and anti-fungal use in hair and skin care products and anti-fungal sprays in agriculture.

8. CLINICAL OUTCOME STUDIES

Three separate studies have been completed in Shanghai. The results of these studies are described below.

8.1 CLINICAL STUDIES ON ALLERDYNE

Allerdyne was used on patients that have been diagnosed as having allergic sinusitis with possibly mild to moderate forms of asthma. Patients with allergic sinusitis took Allerdyne for two months while those that also had asthma received the medication for three months. Data was collected on 55 patients.

The results indicate very favorable results with all patients. Those with asthma and sinusitis also showed positive results though somewhat less so than in the case of those with only allergic sinusitis. There were no significant adverse side effects with any of the patients (Table 3 and 4).

Table 3: Results of all patients who received Allerdyne.

Excellent	65.5%
Good	21.8%
Fair	12.7%
Poor	0%

Table 4: Results of patients with asthma and sinusitis who received Allerdyne

Excellent	52.4%
Good	23.8%
Fair	23.8%
Poor	0%

8.2 CLINICAL STUDIES ON RUBRI-DERM FOR THE TREATMENT OF SKIN DISEASES

A study has been done on **Rubri-Derm** for the treatment of a variety of skin diseases. Out of a group of 80 patients, **Rubri-Derm** was shown to be fully effective on fungal infections, skin rash, and burns. It also provided partial relief for gray toenail and hemorrhoids. Out of the 80 patients enrolled, there were 32 male and 48 female, and ages ranged from 12 to 81.

Table 5: Summary of results of a clinical study on **Rubri-Derm**

Diseases	# of patients	Complete recovery	Partial Recovery	No Improvement
Fungal infection	34	34 (100%)		
Dermatitis (sun light)	1		1	
Dermatitis (Allergy)	2		2	
Dermatitis (nervous system)	1			1
Gray toe nail	5	3 (60%)		2
Burn	2	2 (100%)		
Crack/dryness	3	3 (100%)		
Rash	17	17 (100%)		
Hemorrhoids	6	4 (66%)		2
Heat	1		1	
Psoriasis	3			3
Other skin infections	4		4	

One patient dropped out due to itchiness (possible allergic reaction).

8.3 CLINICAL STUDIES ON ARTHRENE FOR THE TREATMENT OF ARTHRITIS

A study has been begun on Arthrene for the treatment of rheumatoid arthritis. An initial group of 12 patients had arthritis, ranging from mild to severe. The patients started by receiving one capsule of Arthrene twice a day. After the third day, the dosage was increased to 2 capsules each time, twice a day. Treatment lasted one month. Four different parameters were determined: swelling of joints, pain under pressure, morning stiffness and grip force (Table 6). No adverse affects were reported.

Table 6: Initial results on Arthrene for the treatment of Arthritis

Parameter	Result
Joint swelling	Reduced by half (on average)
Pain	Reduced by half (on average)
Morning stiffness	Ranges from slight improvement to complete freedom of movement.
Grip force	Increased in all patients; in some, it increased by two-fold.

9. CONCLUSION

MPT has developed a transPacific strategy to develop a number of product platforms based on TCM. Such platforms allow MPT to develop multiple products from the same formulation; eventually leading to high value FDA approved therapeutics. The strategy combines short-term revenues from sTCM and long-term growth from development of drugs in collaboration with major pharmaceutical companies. MPT's strategy is being implemented through collaboration with both US and Asian research laboratories and companies. Favorable results with clinical outcome studies points out the potential for such a strategy.

(Dr. Robert Yuan is a CEO of Marco Polo Technologies, Inc.)

Chapter 21

COMMERCIALIZATION OF CHINESE HERBAL MEDICINE IN THE GLOBAL MARKET

CHUNG-GUANG SHEN
Vice Chairman and Acting President of Sun Ten Pharmaceutical Co. Ltd.
Taipei, Taiwan, Republic of China

Abstract: Concentrated Chinese herbal extracts have been used for over 40 years in Taiwan, and have been approved for medical reimbursement by the government health insurance system. Recently, the Executive Yuan has directed the Biotech Strategic Review Board (SRB) to include the development of Chinese herbal medicine as one of the national development projects. The Health Administration also issued Guidelines for the Clinical Trial of Chinese Herbal Medicines, undertaking to evaluate the clinical effects of Chinese herbal medicine.

Since the U.S. FDA announced that there will be a the draft of "Guidance on Botanical Drugs Products" in 1997, the scientific modernization and standardization of Chinese herbal medicines has drawn greater international attention and importance. The quality of Chinese herbal medicines is determined by safety, consistency, stability and efficacy. The efficacy is based on clinical validation and evaluations of herbal components; the correct use of herbs and rational chemical evaluations are equally important. Safety, consistency, stability and efficacy are what Sun Ten strive to achieve.

Since Sun Ten began exporting Chinese herbal extracts to Japan in 1975, the herbal extracts have been granted drug reimbursement status by the Japanese health insurance system. Due to strict manufacturing processes and quality control, Sun Ten's herbal extracts have established a very good reputation at home and abroad. Sun Ten is the first TCM manufacturer to receive Taiwan government GMP certification, and also obtained GMP certification by the Australian Therapeutic Goods Administration (TGA), and the Malaysian Health Administration. Sun Ten TCM herbal extracts have been exporting to Japan, Australia, Southeast Asian, the U.S. and various European countries. These herbal extracts have been accepted by the American Special Health Project (ASHP), and have been accepted through contract by other American and Australian private insurance companies. Sun Ten continued to face the challenge by working hard in obtaining ISO 9002 certification in 1999-2000 in recognition of its product quality.

Yuan Lin (ed.), Drug Discovery and Traditional Chinese
Medicine: Science, Regulatory and Globalization, 201-207.
©2001 *Kluwer Academic Publishers. Printed in the Netherlands.*

1. CHINESE HERBAL MEDICINE AND ITS FUTURE IN THE GLOBAL MARKET

Table 1 shows the global market of Chinese herbal medicine in 1997 and projected sales in 2006. The annual growth since in the 1997 in the USA, Germany, France, Vietnam and China, ranges from 8% to the 20%. But in Japan the growth is only 2.6%. In Taiwan, there is a price control by the national health insurance system; the growth only the 4%. The growth will increase more if more traditional Chinese medicine can be exported to the worldwide markets.

Table 1. Sales of herbal medicine in Global market

Area	Sales (1997) USD in billion	Annual growth (%)	Estimated Sales (2006) USD in billion
USA	3.2	15 –20	10.5
Germany	3.8	8	7.5
France	1.8	10	4.2
Mainland China	3.4	8-10	7.0
Japan	2.5	2.6	3.0
Taiwan	0.6	4	0.7
Total	15.3		32.9

Data from: Development Center of Biotechnology, Taiwan

There are some clear difference in the use of western chemical drug and Chinese herbal medicine; i.g. disease targets, mode of action and therapeutic concept. In recent years, there are strong driving forces for the growth of herbal medicine in the global market. The following lists are some of the key documents intended to promote the globalization of herbal medicine.

- WHO (world Health Organization) published a monograph on medicinal plants. WHO and FAO (Food and Agricultural Organization) published a globally common specification of foods.
- The European Union published a monograph on medicinal plants.
- United States Congress passed the Dietary Supplement Health and Education Act (DSHEA) in 1994 and FDA published a draft Guidance on Botanical Drug Product for Industry, August 2000.

There are also many regulatory policy changes in the herbal medicine commercialization in many countries. The regulatory policy is always the key issue to influence the industry in product development. For example, in the United State, the publication of DSHES in 1994 shaped the herbal product in the US market. The publication of a draft Guidance on Botanical Drug Product for Industry this year by FDA outline the requirement to market herbal product as drugs. Following this guidance, botanical products can be developed into drug products by going through New Drug Application process. Many other regions, including Germany, Japan and Mainland China also set up new regulatory policies for herbal medicine in recent years.

2. DEVELOPMENT OF TCM INDUSTRY IN TAIWAN FOR GLOBAL MARKET

Table 2 shows the annual sales of TCM in Taiwan from 1991 to 1998. The product value is quite small: in 1991, only 100 million US dollar; in 1996, 134 million US dollar. From the year of 1996, the national health insurance system was established with a price control, thus the increased is small, in 1997, 130 million; and 1998, 132 millions. The TCM accounts for 11 percent of total market of medicinal product in Taiwan, and 3% of TCM annual sales is from export to Japan, Australia and Malaysia.

Table 2. Annual sales of TCM in Taiwan

Year	Product Value
1991	NTS 3.24 Billion (US $100 Million)
1996	NTS 4.31 Billion (US $134 Million)
1997	NTS 4.18 Billion (US $130 Million)
1998	NTS 4.26 Billion (US $132 Million)

The total sales of Chinese herbs in Taiwan is 640 million in total. But only the 142 million was used for as Chinese medicine. The majority of the herbs were used as healthcare food (47%) and dietary supplements (33%).

The Taiwanese government also actively promoting tradition Chinese medicine industry by setting long-term policy.
Goal:
a. 3-5 products for clinical trial in 3 to 5 years
b. 1-2 products to be marketed as drugs in 5-10 years in the global market.

Funding for R&D from Taiwanese government amounts to US$100 million in 5 years. Most of the funding is for he following 3 institutes:

    a. Pharmaceutical Industry Technology and Development Center-10 years project.
    b. Develop Center for Biotechnology- 5 years project.
    c. Biomedical Engineering Center- 5 years project.

The following flow chart depicts the commercialization of herbal product from research and development (R&D) efforts, particularly pertinent to herbal drug product development.

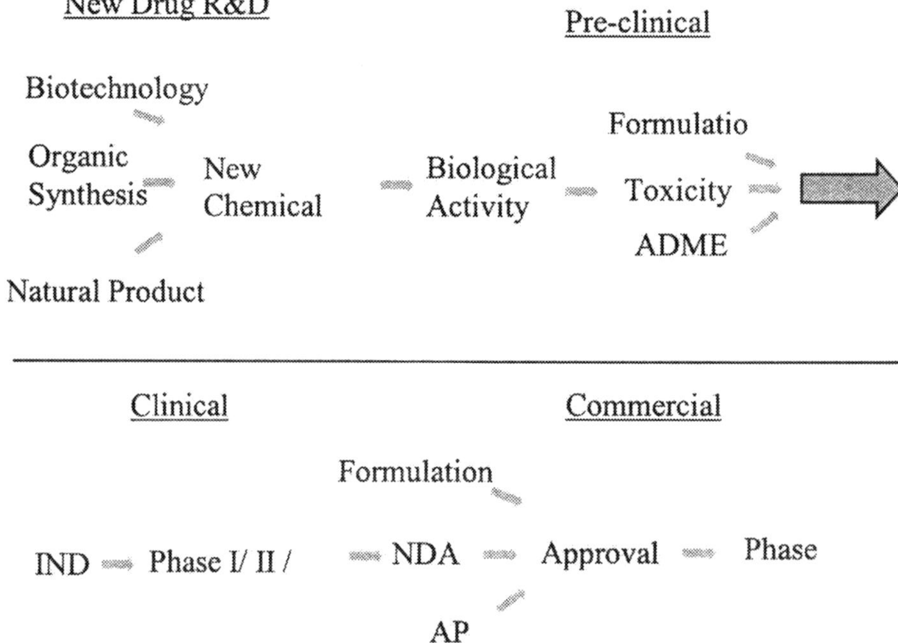

3. THE SUN TEN'S APPROACH

Historical milestone of Sun Ten:

1946   Established Sun Ten Pharmaceutical Company by Dr. Hong Yen Hsu

1975   Exported to Japan; consumer expenses reimbursable from the Japanese Health insurance system

1976   Establish Oriental Healing Arts Institute in Los Angeles, USA

1988   The first TCM manufacturing facility conformed to Taiwanese Good Manufacturing Practice (GMP)

1992   The first company in Taiwan to obtain GMP certification from Australia
1996   Approval for importation by Drug Control Administration of Malaysia
1998   The first National Quality award
1999   Certified ISO 9002 by BSI

Figure 1 below depicts the product development and global marketing strategies of Sun Ten.

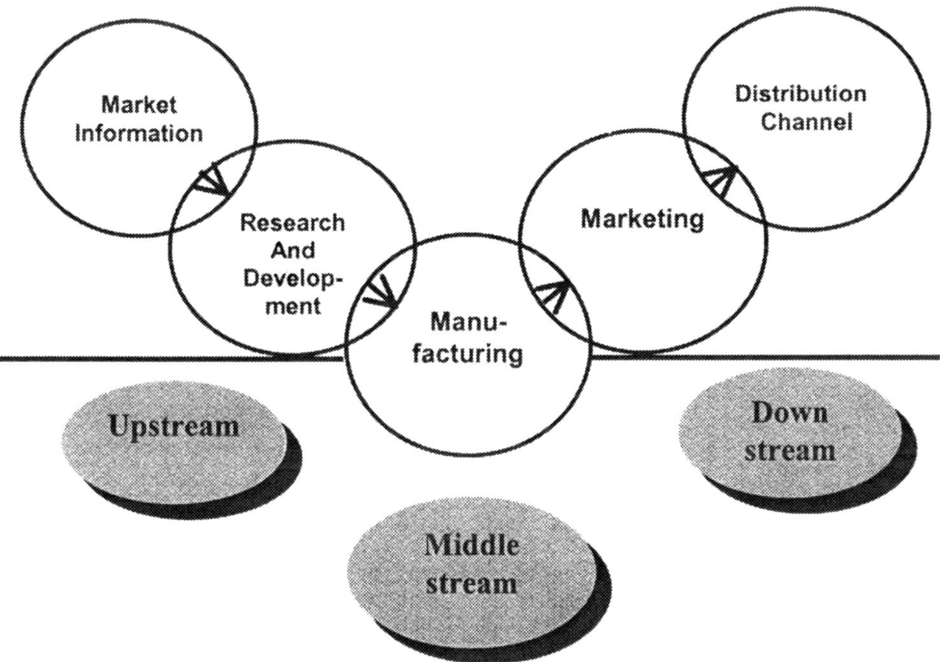

Figure 1. Product Development and Global Marketing Strategies

Sun Ten has its own R & D Center which was established in1972, the Taiwan Brion Research Institute. Raw materials were collected from the harvest area in Mainland China. The original species identification was checked by chemical analysis. The heavy metal content was determined by atomic absorption to meet the domestic requirement. Sun Ten also has set up new analytic technologies for quality control in order to meet the requirement for the market in the United States. Pharmacological properties of products were performed for the IND applications.

Figure 2 showed the high standard of quality control to assure authenticity, safety, efficacy, consistency and stability of Sun Ten' products.

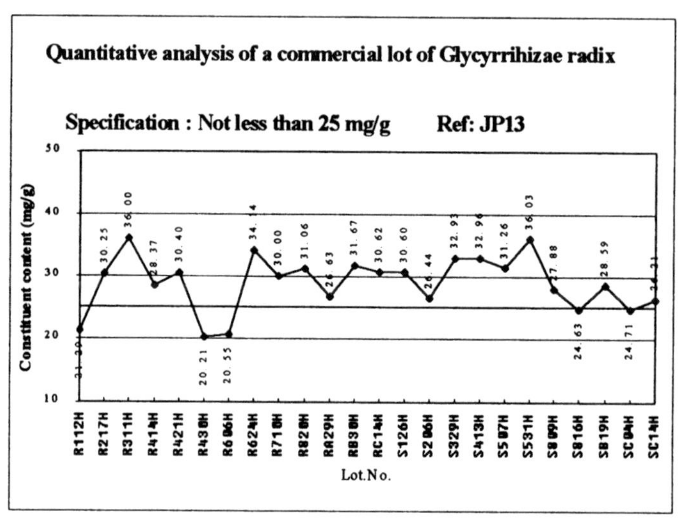

Fig. 2   Quantitative analysis of a commercial lot of Sun Ten's product, Glycyrrihizae radix.

Sun Ten also entered into collaborative research and clinical trials with many other institutes worldwide. Table 3 showed two of our products entering into clinical studies in the US.

Table 3. TCM products for clinical studies in USA

	PHY906	STA-36
Formula	TCM formulation	New TCM formulation
Indication	Cancer therapy	Asthma
Sponsor	PhytoCeutica	Sun Ten, Taiwan
Drug substance manufacture	Sun Ten, Taiwan	Sun Ten, Taiwan
IND submission	2000	2001

4. CONCLUSION

For TCM to enter into the global market, it is the future trend to: (1). promote multiple applications (2). develop multiple dosage forms, (3). develop new Chinese herbal formulas and (4). find new ways for TCM herbal formulas to be integrated into Western medicine. Modern packaging, diverse marketing channels, strategic alliances, and e-commerce opportunities will also be crucial to give TCM a totally different marketing outlook.

Chapter 22

THE DEVELOPMENT OF MODERN TCM IN CHINA TIANFA GROUP

JIALONG GONG
President of Tianfa Co. Ltd, Jinzhou, Hubei, China

Abstract: Traditional Chinese Medicine has been used in China for thousands of years. Hubei province is the hometown of Li Shizheng, the "king of medicine" in ancient China. Now, the Tianfa group of Hubei is actively developing Chinese Medicine. The Tianfa Co. Ltd. is very honored to have had the chance to jointly hold the first International Chinese Medicine Meeting at the University of Maryland (UMD) with the Institute for Global Chinese Affairs (IGCA), UMD, the China Science and Technology Administration, and others. The Tianfa Co. Ltd and its sub institutes are grateful that these organizations came together to discuss Chinese Medicine. Mr. Ma Chengling, the leader of our delegation, has already provided an introduction to the resources and prospects of Chinese Herbal Medicine in Hubei. This paper continues to explain one of the most important companies involved in the development of Chinese Medicine, Tianfa Co. Ltd., which includes the Tianfa Company and the Huoli28 Company. Tianfa Co. Ltd., currently in a time of growth and change, seeks international cooperation in the Hubei Chinese Medicine industry.

1. THE BACKGROUND OF TIANFA CO. LTD.

Tianfa Co. Ltd., founded in 1989, owns Tianfa Company and Huoli28 Company, which are currently both on the market. Large companies, such as Tianrong Modern Agriculture Company and Shuailun Paper Company are also part of Tianfa Co. Ltd. The total assets of Tianfa Co. Ltd. Have reached ten billion RMB, and there are almost ten thousand.

*Yuan Lin (ed.), Drug Discovery and Traditional Chinese
Medicine: Science, Regulatory and Globalization*, 209-211.

As of April 19, 2000, the Tianfa Co. Ltd had bought 45% of the shares of Huoli28 Company. Now, with the support of Tianfa Company, Huoli28 Company is making advancements in the health and high-technology biological medicine fields.

2. THE ADVANTAGE OF DEVELOPING CHINESE MEDICINE IN HUOLI28 COMPANY

As Mr. Ma Chengling explained earlier, the Hubei province is located in the middle of China and has special natural conditions, such as plentiful rainfall, a long frost-free period, plains, mountains, and the unique Shenglongjia Primitive Forest. Hubei is one of the Chinese provinces that is rich in medicine resources.

In addition, there is developed transportation infrastructure. Hubei province is a national traffic pivot, providing various means of water, land, and air travel. These advantages make it the main production and distribution area of Chinese Medicine.

Hubei and its surrounding provinces are densely populated with 1/15 of the world's population living in this area, which makes this region an important medicine market. Hubei province, the birthplace of Mr. Li Shizheng, the famous "king of the medicine" in the ancient China, is now a key location for Chinese Medicine production and distribution. In Hubei, there are about 300,000 medical practitioners and researchers who are devoted to developing the Chinese Medicine resource.

Recently, China has invested more time and resources to the development of Chinese Medicine. In order to take advantage of the possibilities to develop Chinese Medicine, Hubei People's Government has listed Chinese Medicine as one of the emphasis development areas. With support from the provincial Chinese Medicine Research Institute, Wuhan Plant Research Institute of China Academy, Hubei Chinese Medicine College, and others, a number of large-scale medicine production factories have been built. With the publication and implementation of the GAP (Good Agricultural Practices) standard in China, control of the breeding and planting processes, harvesting season and time, primary treatment, and production will be more efficiently controlled. The quality of raw materials and the medicine content will greatly improve.

Huoli28 Company, one of the most famous companies in China, has grasped the great opportunity to develop Chinese Medicine. With the support from its parent company, Tianfa Co. Ltd., Huli28 Company has begun to expand in the field of Chinese Medicine.

3. THE HUOLI28 COMPANY DEVELOPING CHINESE MEDICINE

In the 21st Century, the development of the international medicine industry requires combined efforts from various disciplines, including Biological Engineering, Synthetic Medicine, and Natural Plant Medicine. Huoli28 Company will emphasize the scale and industrial production of Chinese Medicine and plant extraction. Now that the company has bought 560,000 square meters of land in Jingzhou City, Hubei province, there are plans to build a large facility, known as the "Huoli Medicine and New High-Technology Biology Garden." The facility will also include an auxiliary pharmacy production base, exchange market, and means of Internet access for electronic business. The endeavor will create an international production, supply, and sale system. Additional goals include that the company plans to jointly build a national Chinese medicine development and research institute with the cooperation of related organizations. Huoli28 Company will also build an international standard for similar biological engineering projects.

The planning and implementation of "Huoli Medicine and New and High Technology Biology Garden" will be a starting point for the development of Hubei Chinese Medicine. There is no doubt that Huoli28 Company will play an active role in promoting the incorporation of Chinese Medicine into the world market. Tianfa Co. Ltd. welcomes other countries to cooperate with us and jointly build the "Huoli Medicine and New and High Technology Biology Garden."

FDA FORUM

FDA forum was presented by a panel of experts from FDA and pharmaceutical industry. The forum was chaired by Dr. Yuan Yuan Chiu. Each panelist presented their view on issues relating to the regulation of herbal product in the U.S., either as a drug or as a dietary supplement. The overhead in the following pages are their contributions.

PANEL MEMBERS AND THEIR AFFILIATIONS:

CHIU, Yuan Yuan   Director, Office of New Drug Chemistry
Center for Drug Evaluation and Research
U.S. Food and Drug Administration

CHEN, Chi-Wan   Director, Division of New drug Chemistry
Center for Drug Evaluation and Research
U.S. Food and Drug Administration

CHIN, Ling   Medical Officer, Division of Over-the –
Counter Drug Evaluation
Center for Drug Evaluation and Research
U.S. Food and Drug Administration

LOVE, Lori A   Director, Clinical Research and Review
Office of Nutritional Products, Labeling and
Dietary Supplements, HFS 805
Center for Food Safety and Nutrition
Food and Drug Administration

PAGE, Samuel W.   Director for Research
Joint Institute for Food Safety and Applied
Nutrition

BETZ, Joseph, M: Vice President
American Herbal Products Association

HOFFMAN, Freddie Ann   Senior Director for Medical and Clinical
Affairs, Pfizer

1. REGULATING BOTANICAL PRODUCTS: A FDA OVERVIEW

YUAN YUAN CHIU
Director, Office of New Chemistry Center for Drug Evaluation and Research
Food and Drug Administration

Unique Issues Related to Botanical Products in US

- The enactment of Dietary Supplements, Health and Education Act in 1994
  - Drugs or Dietary Supplements?
- The existence of present or past human experiences in USA or foreign countries
  - New drugs or GRASE (generally recognized as safe and effective) drugs?
- The products contain multiple components from single, or multiple plants
  - Apply Fixed-Combination Drug Regulations or not?

Drugs or Dietary Supplements? Why Does It Matter?

- Drugs
  - Prove to be safe and efficacious for its intended use by manufacturer
  - Require prior approval by FDA before marketing
    - Product quality, safety and efficacy reviewed
  - Require FDA approval before conducting human study for non-approved drugs/uses
    - Product quality, safety, and study design reviewed

Drugs or Dietary Supplements? Why Does It Matter?

- Dietary Supplements
  - Consider to be food - assumed safe
    - Burden of proof on FDA
  - Labeled statements on the effect(s) of the product should be substantiated
    - FDA has no legal authority to evaluate the validity of the statement
  - Require no FDA approval before marketing
    - Disclaimer required
  - Require no FDA approval before conducting human study

Drugs or Dietary Supplements? How Are They Defined?

- Based on "intended use/labeled claim"
- Food, Drug and Cosmetic Act
  - Drugs: An article used to diagnose, cure, mitigate, treat or prevent disease
  - Drugs: An article (other than food) used to affect the structure or function of body

Drugs or Dietary Supplements? How Are They Defined?

- Dietary Supplement, Health, and Education Act (Nov., 1994) - Dietary Supplements
  - An herb or other botanical (other than tobacco)...
  - Concentrate, metabolite, constituent, extract, or combination of
  - Intended as a supplement to the diet by ingestion
  - Not containing an approved drug
  - Permitted labeled claims - effects on structure and function of body

Regulations for Dietary Supplements Structure or Function (21CFR101.93)

- Final rule issued Jan. 6, 2000
- Effective date February 7, 2000
- 21CFR101.93(f)
  - *Permitted structure/function statements*
    - ...to affect the structure or function in human...provided the statements are not disease claims

Regulations for Dietary Supplements Structure or Function (21CFR101.93)

- 21CFR101.93(g)
  - *Disease claims*
    - a "disease" is damage to an organ, part, structure, or system of the body such that it does not function properly... or a state of health leading to such dysfunctioning..., except that diseases resulting from essential nutrient deficiencies...
    - Abnormal conditions associated with a natural states or processes, if the abnormal condition is uncommon or can cause significant or permanent harm
    - Labeling restrictions

Regulating Dietary Supplements

- Proposed Rule on labeling
- Advance Notice of Proposed Rule on GMP
- Safety evaluation
- Quality issues

New Drugs or GRASE Drugs? Why Does it Matter?

- GRASE Drugs
  - OTC Monograph Petition
  - NDA Approval
    - OTC dispensing
    - Rx dispensing
- New drugs - Not GRASE
  - IND (for US studies)/NDA Approval
    - OTC dispensing
    - Rx dispensing
- Same evidence required to show S/E

OTC Monograph Drugs

- Current rules: 21CFR Part 330
- Proposed rule (Dec. 20, 1999): Additional Criteria and Procedures for Classifying OTC Drugs as GRASE and Not Misbranded (foreign data)

Draft Guidance for Industry: Botanical Drug Products (Issued on 8/10/00)

- Scope
  - Inclusion - Crude extracts of plant(s), alga(e) or macroscopic fungus(i)
  - Exclusion - Highly purified, chemically modified or semi-synthetic compounds from plants

---

Draft Botanical Drugs Guidance

- Regulatory Issues
  - OTC Monograph/New Drugs
  - <u>IND submission required for human studies on non-approved botanical drugs if conducted in USA for research purpose or for supporting commercialization</u>
  - Combination Drug Regulations
    - Complex mixtures
- Scientific/Technical Issues
  - General Principles
  - Focused on INDs
  - Standards for S/E remain unchanged

Regulations on Fixed-Combination Drugs Rx (21CFR300.50)

- Two or more drugs may be combined in a single dosage form when each component makes a contribution to the claimed effects and the dosage of each component (amount, frequency, duration) is such that the combination is safe and effective…

Regulations on Fixed-Combination Drugs OTC (21CFR 330.10(a)(4)(iv))

- An OTC drug may combine two or more safe and effective active ingredients and may be generally recognized as safe and effective when each active ingredient makes a contribution to the claimed effect; when combining of the active ingredients does not decrease the safety or effectiveness of any of the individual active ingredients...

Draft Botanical Drugs Guidance

- Products derived from a single part of a plant, or from an alga or macroscopic fungus, are not considered to be fixed combination drugs within the meaning of 21 CFR 300.50 and 330.10(a)(4)(iv)

Regulations : Fixed-Combination Drugs Requirements (To-be-Proposed)

- Combination drug requirements (each contribution to claimed effects, etc.) essentially unchanged
- Categorical exemption: Drugs derived from single animal, or botanical raw material (e.g., a single part of a plant)
- Conditional exemption: Drugs derived from multiple natural-sources when certain conditions are met
- General waiver provision

Botanical Products #1

Dietary Supplements	Drugs
• Food	• OTC or Rx drugs
• Ingested	• Any routes of administration
• Food (Dietary Supplement) GMP	• Drug GMP
• Structure or function claims	• Disease claims and/or structure or function claims
• FDA to prove unsafe	• Safety proven
• Efficacy not a concern	• Efficacy proven
• No marketing approval from FDA	• FDA approval prior to marketing

Botanical Products #2

Dietary Supplements	Drugs
• Authenticity needed	• Authenticity needed
• Potency not an issue	• Potency critical
• Assay of markers adequate	• Assay of markers not adequate unless clinical relevant
• Batch to batch (non-marker components) consistency not critical	• Batch to batch (whole mixture) consistency critical
• Equivalence among products of different firms not an issue	• Equivalence among products of different firms not assumed

Summary

- FDA is revising regulations to meet the unique needs of botanical products
- FDA is providing guidance on studying botanical drugs based on science
- FDA welcomes applicants to submit NDAs or OTC monograph petitions
- FDA encourages sponsors to submit INDs
- FDA is willing to work with sponsors
- FDA review staff are available for pre-submission consultations

2. DRAFT FDA GUIDANCE ON BOTANICAL DRUG PRODUCTS—FOR INDS

CHI-WAN CHEN
Office of New Drug Chemistry Center for Drug Evaluation and Research
U.S. Food and Drug Administration

Draft Guidance on Botanical Drug Products

- Published 8/10/00 for public comment
- www/fda.gov/cder/guidance/1221dft.pdf

Guidance on Botanical INDs

- IND regulation
- Phase 1/2 clinical studies of lawfully marketed botanical products
- Phase 1/2 clinical studies of non-marketed botanical products
- Phase 3 clinical studies of all botanical products

IND Regulation

- Amount of information required depends on:
  - Novelty of the drug
  - Extent to which the drug has been studied previously
  - Known or suspected risks of the drug
  - Developmental phase of the drug

IND Regulation (cont'd)

- Format
  - Description of product and documentation of human use
  - Chemistry, manufacturing, and controls (CMC)
  - Pharmacology/toxicology
  - Bioavailability
  - Clinical

General CMC Principles for Botanical INDs

- No need to identify or chemically assay actives or markers if not feasible
- Chromatographic fingerprint and strength by weight acceptable
- Adequate raw material controls and in-process controls for drug substance emphasized
- Sufficient quantities from a single batch and single source recommended

---

Phase 1/2 for Lawfully Marketed Botanical Products

- CMC
  - Raw material
    - Brief description
    - Certificate of authenticity if marketed outside U.S.
  - Drug substance
    - Type of process

Phase 1/2 for Lawfully Marketed Botanical Products (cont'd)

- CMC (cont'd)
  - Drug product
    - Qualitative/quantitative description; no adulteration
    - Manufacturer's certificate of analysis if available
    - Quality testing if COA not available for product marketed outside U.S.

Phase 1/2 for Lawfully Marketed Botanical Products (cont'd)

- CMC (cont'd)
  - Placebo
  - Labeling
  - Environmental assessment or claim of categorical exclusion

---

Phase 1/2 for Lawfully Marketed Botanical Products (cont'd)

- Pharmacology/toxicology
  - Previous human experience acceptable
  - Animal toxicity data if available
  - Data to support safe human use if marketed outside the U.S.

Phase 1/2 for Lawfully Marketed Botanical Products (cont'd)

- Bioavailability
  - Standard PK/PD may not be feasible unless actives or markers are known
- Clinical
  - Well-controlled study capable of demonstrating effectiveness recommended
  - Randomized, parallel, dose-response study if dose is in doubt

---

Phase 1/2 for non-Marketed Botanical Products

- CMC
  - Raw material
    - Description
    - Certificate of authenticity
    - Grower/supplier
    - Harvest location/time
    - Collection and preservation procedures
    - Storage

Phase 1/2 for non-Marketed Botanical Products (cont'd)

- CMC (cont'd)
  - Drug substance
    - Qualitative/quantitative description
    - Manufacturer
    - Type of process
    - Quality control tests, including identity by fingerprint, strength by weight, heavy metals, microbial limits, and, if feasible or potent substance, chemical assay, bioassay

Phase 1/2 for non-Marketed Botanical Products (cont'd)

- CMC (cont'd)
  - Drug product
    - Qualitative/quantitative description; no adulteration
    - Manufacturer
    - Manufacturing process
    - Quality control tests, including identity by fingerprint, strength by weight, microbial limits, dosage-form-specific tests, and, if feasible or potent substance, chemical assay, bioassay

Phase 1/2 for non-Marketed Botanical Products (cont'd)

- CMC (cont'd)
  - Drug product (cont'd)
    - Container closure
    - Stability data
  - Placebo
  - Labeling
  - Environmental assessment or claim of categorical exclusion

Phase 1/2 for non-Marketed Botanical Products (cont'd)

- Pharmacology/toxicology
  - Traditional preparations: Previous human experience may be sufficient without standard preclinical testing
  - Non-traditional preparations: Preclinical testing needed depends on indication, extent of safe human experience, safety concerns about new formulation, processing, etc.

Phase 1/2 for non-Marketed Botanical Products (cont'd)

- Bioavailability
  - Blood levels of known actives or markers if feasible
  - Drug-drug interaction if co-administered with other drugs
- Clinical
  - Same as for marketed botanical products; but greater assurance of safety needed

Phase 3 for All Botanical Products

- CMC (in addition to CMC for Phase 1/2 non-marketed botanical)
  - Raw material
    - More details on manufacturing process
    - Quality control tests, including chromatographic fingerprint, chemical identity of actives or markers, heavy metals, microbial limits, residual pesticides, adventitious toxins, foreign materials, and bioassay if available
    - Reference standard

Phase 3 for All Botanical Products (cont'd)

- CMC (in addition to CMC for Phase 1/2 non-marketed botanical) (cont'd)
  - Drug substance
    - Chemical identity of actives or markers
    - More details on process
    - More quality control tests, including chemical ID, residual pesticides, adventitious or endogenous toxins
    - Analytical procedures
    - Reference standard
    - Batch analysis
    - Container closure
    - Stability data

Phase 3 for All Botanical Products
(cont'd)

- CMC (in addition to CMC for Phase 1/2 non-marketed botanical) (cont'd)
  - Drug substance
    - Chemical identity of actives or markers
    - More details on process
    - More quality control tests, including chemical ID, residual pesticides, adventitious or endogenous toxins
    - Analytical procedures
    - Reference standard
    - Batch analysis
    - Container closure
    - Stability data

Phase 3 for All Botanical Products
(cont'd)

- CMC (pre-NDA objectives)
  - Adequate controls for raw materials
  - Validated process and adequate in-process controls
  - Batch-to-batch consistency based on fingerprinting
  - Adequate specifications, including ID and assay for actives, ID and assay for marker, and/or bioassay
  - Validated analytical procedures
  - Suitable reference standards

Phase 3 for All Botanical Products
(cont'd)

- CMC (pre-NDA objectives) (cont'd)
  - Stability-indicating analytical procedures for stability studies
  - Comparison among preclinical, clinical, and to-be marketed products
  - CGMP conformance for manufacturing and testing facilities
  - Environmental assessment or claim of categorical exclusion

Phase 3 for All Botanical Products
(cont'd)

- Pharmacology/toxicology - Phase 3 and pre-NDA
  - Previous human experience may be insufficient, especially for chronic therapy
  - Standard preclinical studies, e.g., general toxicity, teratogenicity, mutagenicity, and carcinogenicity, may be needed

Phase 3 for All Botanical Products
(cont'd)

- Bioavailability
  - If feasible, blood levels of known actives, markers, or active metabolites, or pharmacological effect using bioassay; if not, clinical effects observed in well-controlled trials
  - Drug-drug interaction
  - Effect of impaired clearance
  - Linkage between different strengths
- Clinical: Same as for synthetic drugs

3. OTC DRUG TO MARKET

LING CHIN
Food and Drug Administration Center for Drug Evaluation and Research
Division of Over The Counter Drug Products

OTC Drugs in Market

- **Two basic ways:**
  - OTC Drug Rview
  - New Drug Application

Monograph Method

- **To remain on the market**
  - OTC Drug Review
- **To get to market**
  - Switch regulation
  - Panel recommendation

OTC Drug Review

- **1972: OTC Drug Review:**
  - Ingredients by therapeutic category.
    e.g. Psyllium as a bulk laxative
  - Monograph: GRAS and GRAE
  - Non-Monograph: failure to demonstrate
    safety and efficacy

OTC Drug Review

- **Three phase rulemaking process:**
  - Each phase announced in Federal Register
- **Phase 1 (ANPR)**
  - FDA-appointed advisory review panels
  - **Category I:** GRASE
  - **Category II:** NRASE
  - **Category III:** insufficient data

OTC Drug Review

- **Phase II (PR)**
  - FDA Review: panel's findings, public
    comments, new data, etc
- **Phase III (FR)**
  - Further FDA review: public comment, new
    data, hearings, etc
  - **Final Monograph**
  - **Final Non-Monograph**

Botanicals

- **Already reviewed:**
  - 21% of botanical mentions
- **Final Monographs:**
  - Category I: 3%
  - Category II: 71%
- **Pending review:**
  - 24% of botanical mentions

Monographs

❖**Anorectal:** cocoa butter: external protectant
❖**Antacid:** tartaric acid
❖**Cough Cold:** camphor: antitussive
                peppermint oil: antitussive
❖**Skin Protectant:** witch hazel (astringent)

Prescription Vs. OTC

1951: Durham-Humphrey Amendment:

- 3 classes of drugs limited to Rx:
  - certain habit-forming drugs listed by name in the Food, Drug & Cosmetic Act
  - drugs not safe for use except under supervision of licensed practitioner
  - drugs limited to Rx under an NDA

Prescription Vs. OTC

- Requiring supervision for reasons of:
  - toxicity
  - other potential for harmful effect
  - method of use (e.g. IV)
  - collateral measures necessary to use (e.g. labs)

NDA Method

- **To get to market**
  - Rx to OTC
  - Direct to OTC

Rx-to-OTC

- Switched drugs:
  - known drugs
  - extensive experience
  - same indication, same dose
  - different indication, different dose

OTC Studies

- Label Comprehension Studies
- Actual Use Studies
- Adequate and well controlled
- Evidence to demonstrate that adequate directions can be written for the safe and effective use of the product by consumers

Level of Evidence

- Label Comprehension Studies
  - label and/or package insert tested
  - no actual use of drug
  - demonstrate label adequate for appropriate self-selection and use

Level of Evidence

- Actual Use Studies
  - self-selection
  - drug dispensed
  - behaviour monitored
  - demonstrate use without learned intermediary is safe and effective

4. DIETARY SUPPLEMENT PRODUCTS: SAFETY AND REGULATORY CONCERNS

LORI A. LOVE
*Director, Clinical Research and Review Center for Food Safety and Applied Nutrition
Food and Drug Administration*

D ietary
S upplement
H ealth and
E ducation
A ct

Dietary Supplements: DSHEA Definition

Product (other than tobacco)
Intended to supplement the diet
Contains one or more of the following dietary ingredients:
  A vitamin
  A mineral
An herb or other botanical (not tobacco)
  An amino acid
  Any other "dietary substance" for use by man to supplement the diet by increasing the total dietary intake
  A concentrate, metabolite, constituent, extract, or combination of any of the above

---

Dietary Supplements under DSHEA

- ingested product that is:
  - In tablet, capsule, liquid, powder, gelcap, softgel
  - Not represented as conventional food
  - Not represented as sole item of meal
  - Not represented as a total diet
  - Labeled as a dietary supplement
- Can include articles previously approved as a drug, antibiotic, or biologic, if first marketed as a dietary supplement
- Will not generally include articles approved, or authorized for investigation, as a new drug, antibiotic, or biologic that were not first marketed as a dietary supplement

---

Dietary Supplements

- No pre-market registration, review, or approval by FDA
- Exempt from food additive provisions
- "Optional" GMP regulations

Safety Standards

• All Other Foods	• Dietary Supplements
Demonstrate safety – "intended conditions" – GRAS – Food Additive	– Significant or unreasonable risk of illness or injury under labeled conditions – Adulterated if contains "new dietary ingredient" without 75-day notification (1997 regulation)

Why historical use can not be relied on to provide evidence of safety:

- No documented systems to collect and evaluate adverse effects associated with product use
- Different products, populations and use patterns today compared to historical use

Dietary Supplement Safety

- How does FDA learn about potential safety problems with dietary supplements?

- DSHEA Places the Burden of Proof on FDA

FDA Reporting Systems for Botanicals

- No central system based on type of ingredients
  - Adverse event report goes ultimately to Center with regulatory responsibility for the particular product

Time Trends on Adverse Events

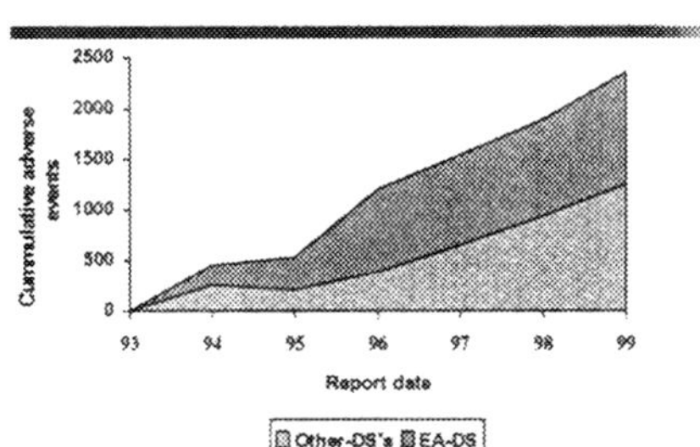

Dietary Supplements with Safety Concerns:

Product / Ingredient	Adverse Effects
Vitamin A	Teratogenecity, hepatotoxicity
Niacin	Hepatotoxicity, myopathy
L-Tryptophan/ 5-HTP	Eosinophilia myalgia syndrome
GBL/GHB	CNS depressant

Hormones:
  melatonin
  androstenedione
  DHEA
	??

Dietary Supplements with Safety Concerns:

Product / Ingredient	Adverse Effects
Ephedra spp. (ma huang)	CVS and CNS
Germanium	Nephrotoxicity
Chaparral (Larrea dentata)	Hepatotoxicity
Comfrey (Symphytum ssp)	Hepatotoxicity
Germander (Teucrium chamaedrys)	Hepatotoxicity, CNS

Safety Considerations

- Product is directly harmful
- Product is adulterated or contaminated
- Product - product or other co-factor interactions occur
- Product is substituted for a known effective therapy

Safety Concerns Because of Contaminants or Adulterants

- Pesticides
- Microbial contamination
- Molds, mycotoxins
- Filth
- Heavy metals:
  - Bee products contaminated with lead
- Drugs, chemicals
  - Black pearls, jin bu huan other "patent" medicines
- Misidentified or substituted ingredients
  - Plaintain contaminated with digitalis

Product Interactions

- Increased anticoagulant effects
  - NSAIDS or warfarin with white willow, garlic, ginger, feverfew or vitamin E
- Increased cardiovascular and nervous system stimulation
  - Cardiac drugs, MAOI, caffeine, or cough/cold products with ephedra, kola, guarana, khat, or yohimbe
  - MAOI, SSRI, or ß-sympathomimetics with St. John's wort

5. JOINT INSTITUTE FOR FOOD SAFETY AND APPLIED NUTRITION (JFSAN)

SAMUEL W. PAGE
Director for Research JFSPAN

<table>
<tr><td>

Why are we (CFSAN/JIFSAN) interested in TCMs?

- Growing consumer interest
- Marketed as dietary supplements
- Safety issues
- Provide technical support for review functions of FDA

</td><td>

Top Selling Herbal Products in the U.S. also Used in TCM

- Ginkgo
- Garlic
- Ginseng
- Ma Huang

</td></tr>
<tr><td>

Problems with Traditional Chinese Medicines

- Lack of significant history of use in U.S.
- Not used according to traditional practices
- Lack of standardized formulations
- Misidentification of plant materials

</td><td>

Problems with Traditional Chinese Medicines

- Illegal addition of prescription drugs and toxic substances
- Inherent toxicity of some of the ingredients
- Materials of animal origin
- Plant/Drug Interactions

</td></tr>
<tr><td>

Problems with Traditional Chinese Medicines

- Our general lack of expertise and knowledge about TCMs

</td><td>

Traditional Chinese Medicines

- Documented existence for over 2000 years
- *Classic of Materia Medica*
  - 252 botanicals
  - 45 minerals
  - 67 zoologicals

</td></tr>
</table>

Chinese Concept of Yang-Yin

- Philosophy was used as a guide for all aspects of Chinese life
- TCMs were not claimed to cure anything, but to support and assist Nature in healing
- No distinction drawn between food and medicine
  - Many of the same botanicals used for medicines and for supplementing and flavoring daily meals

Traditional Chinese Medicines

- Composed of a multitude of ingredients whose interactions with the body are exceedingly complex

Traditional Chinese Medicine

- Combined in prescriptions containing 6 - 12 substances
  - One principle component
  - Several assistant components
- Some exceptions, such as ginseng root and garlic

Addition of Non-declared Drugs/Toxic Substances

- diazepam
- mercury
- cannabis
- codeine
- opium
- prednisolone
- aminopyrine
- phenlybutazone
- arsenic disulfide
- chloramphenicol
- methyltestosterone
- chlordiazepoxide
- tetrahydropalmatine
- methyl salicylate

•Misidentification of Plant Materials: *Aristolochia*

•Inherent Toxicity of TCM Materials: *Aristolochia*, *Aconitum* *Bufo vulgaris*

•Use of Materials of Animal Origin

Transmissible spongiform encephalopathies
New/unrecognized infectious agents

•Plant/Drug Interactions

Factors that influence pharmaco/toxicokinetics

•New Issue

Functional foods based on TCMs

6. INDUSTRY INITIATIVES

Joseph M. Betz
American Herbal Products Association

Initiatives

➢ Herbs of Commerce
➢ cGMP
➢ Adulteration Manual
➢ Botanical Safety Handbook
➢ Trade Recommendations and guidelines
➢ INA/MVP
➢ American Herbal Pharmacopoeia

Dietary Supplement Label

➢ Same as food label **PLUS:**
➢ The words "dietary supplement"
➢ Name and quantity of each ingredient
  - Total weight of each ingredient
  - Plant part used
  - Etc.

Naming Systems

➢ Plants are not limited by national or linguistic boundaries
➢ Common names
  - Same name may be applied to different plants in different parts of a country
  - Same plant may have several common names
  - Some plants have no common name at all

"Drug Names"

➢ **NOT scientific names**
➢ **Found in PPRC, JP, others**
  - *Cimicifugae Rhizoma* (black cohosh)
  - *Glycyrrhizae Radix* (licorice root)
    - *Glycyrrhiza uralensis* Fisher
    - *G. glabra* L.

Scientific Names

➢ There is only one correct scientific name for a plant at any time. There may be some accepted synonyms.

Herbs of Commerce

➢ Marketplace attuned to common names
➢ American Herbal Products Association produced a list of "standardized" common names, including *pin yin* transliterations
  - Assigned a **single, unique** common name to a single correct scientific name
  - HOC ('92) codified in 21 CFR 101.4 (h) as *the* source for common names

Dietary Supplement Regulations

> Advanced Notice of Proposed Rulemaking (ANPR) on Current Good Manufacturing Practice (cGMP) for DS published in 1997 (62 FR 5699-5709)
> • http://vm.cfsan.fda.gov/lrd/fr970206.html
> Does NOT have the force of law, many companies using as guidance

Preventive Measures

> Identification
> - Names
> - Voucher Trained Collectors/Cultivation
> - Correct Specimens/reference material
> Testing
> - Organoleptic/sensory evaluation
> - Microscopic evaluation
> - Chemical Evaluation
> - Raw Materials and finished product

Adulterants

> AHPA Botanical Raw Materials Committee
> *Botanical Adulteration Manual*
> • Volume 1-safety
> • Volume 2-economic

Safety

> **Circumstances of Intoxication**
> - Misidentification
> - Misuse
> - Deliberate Adulteration
> - *Inherent Toxicity*

Safety

> AHPA's Standards Committee created a special subcommittee to address botanical safety issues
> Literature fragmented and incomplete for herbs in North American Commerce
> Commissioned the *Botanical Safety Handbook*
> *>600 plants*

Classification System

The subcommittee placed herbs into safety Classes
> Class 1
> - Herbs that can be safely consumed when used appropriately
>   • *Avena sativa*
>   • *Echinacea* spp.
>   • *Valeriana* spp.
>   • *Urtica dioica*

Classification System

> Class 2
> Herbs for which the following use restrictions apply, unless otherwise directed by an expert qualified in the use of the described substance:
> - 2a: For external use only (comfrey).
> - 2b: Not to be used during pregnancy (dried ginger).
> - 2c: Not to be used while nursing (garlic)
> - 2d: Other specific use restrictions as noted.

Classification System

> Class 3
> - Herbs for which significant data exist to recommend the following labeling:
>   • "To be used only under the supervision of an expert qualified in the appropriate use of this substance."
> - Labeling must include proper use information: dosage, contraindications, potential adverse effects and drug interactions, and any other relevant information

Classification System

> **Class 4**
> - Herbs for which there is insufficient data for classification.
>   - *Uncaria tomentosa*

Trade Recommendations

> **Ephedra** (adopted 3/94; revised 1/2000)
> - **AHPA recommends criteria for marketing of dietary supplements containing ephedrine alkaloids:**
>   - *Labeling*
>   - *Serving Limits*
>   - *Herbs of Commerce Conformity*
>   - *Synthetic Ingredients*
>   - *Marketing*

Botanical Safety Handbook
(adopted 4/98)

> AHPA recommends that any products that contain herbs classified in the *Botanical Safety Handbook* in Class 2b or 2c be labeled according to the labeling classification for those classes.

Pesticide Analysis for Ginseng
(adopted 11/99)

> AHPA recommends that processor and manufacturer members analyze cultivated ginseng (*Panax* spp.) for the presence of quintozene and related compounds by a validated methodology

Known Adulterants
(adopted 4/97)

> AHPA recommends that appropriate steps be taken to assure that the following raw materials are free of the noted adulterant:

Known Adulterants

Herb in Commerce	Adulterant
• Siberian Ginseng root (*Eleutherococcus senticosus*)	• *Periploca sepium* root
• Plantain leaf (*Plantago lanceolata*)	• *Digitalis lanata* leaf
• Skullcap herb (*Scutellaria lateriflora*)	• Germander herb (*Teucrium chamaedrys*)
• Stephania root (*Stephania tetranda*)	• *Aristolochia fangchi* root

Methods Development

> Institute for Nutraceutical Advancement (INA) Methods Validation Program
> - Develops, validates, publishes analytical testing methods for botanical raw materials
> - Industrial sponsors vote on methods, provide technical and/or financial support
> - Submit validated methods to USP and AOACI

American Herbal Pharmacopoeia and Therapeutic Compendium

- Non-profit corporation 501C(3)
- To provide information necessary for quality assurance and safe and effective use of botanical medicines.
- Production of peer reviewed monographs.
- Synthesis of authoritative traditional and scientific literature.
- Reviewed by multidisciplinary committee of medicinal plant experts.
- Goal of developing 300 monographs.

7. U.S. REGULATION OF BOTANICALS AND TCM WHAT NEXT?

FREDDIE ANN HOFFMAN
Warner-Lambert Consumer Healthcare Pfizer

Botanicals in the US
Before 1906

- US drugs *were* botanicals
- 1840's US Pharmacopeia - botanical standards
- immigrants left their "roots" at the border
- unique practice of Medicine in US
- self-medication
- early 20th Century: "Patent Medicines"

Botanicals
US: Changing Landscape 1980's

- multicultural nation
- global communication (Internet, TV, travel)
- globalization of trade
- globalization of pharmaceutical/food industries
- dissatisfaction w/ conventional medicine:
  - rising healthcare costs
  - demand for choice/control issues

US Legislation

- 1990 - *Nutrition Labeling & Education Act*
- 1992 - Congress establishes NIH Office of Alternative Medicine
- 1994 - *Dietary Supplement Health & Education Act*

Regulation of Botanicals
US Agencies

- **Food & Drug Administration**
- Federal Trade Commission
- Environmental Protection Agency
- US Department of Agriculture
- Drug Enforcement Administration
- Fish & Wildlife Service
- US Customs
- Alcohol, Tobacco & Firearms

Principles of US Regulation

- Regulated by "Intended Use" *not by source or origin*
- *Claims* define the Intended Use
- Formulation and safety may further define the regulatory route

---

Products are regulated by their
Intended Use

which is determined by the
Labeling
(claims, advertising, promotion, etc.)

The Regulatory Continuum
Adapted From Jim Lassiter - Nutrilite , A Division of Amway Corp

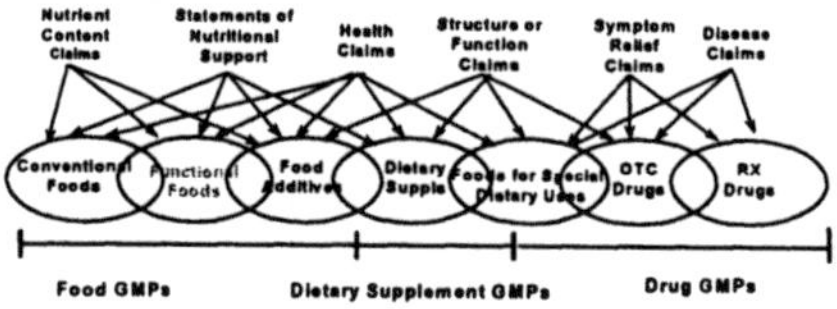

The Future?

Foods <<<< >>>> Drugs

Opposing Legal Premises

Foods
are

- Generally Recognized As Safe ("GRAS")
- Not approved for Safety or Efficacy
- not intended to diagnose, prevent, treat, mitigate or cure disease.

Drugs
are

"....not generally recognized as safe and effective under the conditions prescribed, recommended or suggested in the labeling" ("New Drugs" marketed after 1938)

Drugs
are

- approved for Safety and Efficacy based upon "adequate and well-controlled trials"
- are intended to diagnose, prevent, treat, mitigate or cure disease.

Dietary Supplements (DSHEA):
Implementation

- Labeling regulations
  - **Final Rule 9/23/97**
  - **Enacted 3/23/99** - Compliance actions-YES
- Good Manufacturing Practices (GMPs)
  - **ANPR 2/06/97**
  - **Public Meeting 7/12/99** - Compliance actions??
- "Structure" or "Function" Claims
  - **Final Rule 1/02/00**
  - **Enacted immediately** - Compliance actions-YES

US Regulation - Botanicals
Draft Botanical Drug Guidance
(FR 8/10/00)

- single IND for multicomponent product
- combination Rule - modified
- recognizes "process control" (nongeneric)
- QAQC - drug GMPs
- "history of use"
- advancement of Clinical Program - Phase II/III vs Phase I

Botanicals:
A new class of drugs

- often naturally derived products
- mixtures; heterogeneous
- active principles not always known or defined
- scale up required
- lot release protocols
- defined by process control

Good Agricultural Practices

OTC Monograph

Rx & OTC via IND-NDA

Botanical Standards Monographs

Evolving Marketplace

World Marketplace

Globalization

Forecast for US Botanicals

THE FIRST INTERNATIONAL CONFERENCE ON

TRADITIONAL CHINESE MEDICINE: SCIENCE, REGULATION AND GLOBALIZATION

* * *

PROGRAM

* * *

August 30 - September 2, 2000
Colony Ballroom
Stamp Student Union
University of Maryland, College Park

WEDNESDAY, AUGUST 30, 2000

8:00 am Registration

9:00 am Welcome and Opening Remarks
Dr. William Tai (Director, Institute for Global Chinese Affairs)
Dr. Daniel Mote (President, University of Maryland, College Park)

9:20 am Keynote Lecture
Approaching Traditional Chinese Medicine: Inheritance and Exploration
Professor Yongzheng Hui, Former Vice Minister
(Ministry of Science and Technology, China)

10:00 am Keynote Lecture
Floras, Plant conservation and China's Future
Dr. Peter H. Raven, Director
(Missouri Botanical Garden, Home Secretary and Member of U.S. National Academy of Sciences)

10:40 am Coffee Break

236

Natural Resources and Conservation of Medicinal Plant Species
Co-chairs: Dr. Gordon Cragg (National Cancer Institute, NIH)

Professor De-Yuan Hong (Academia Sinica, China)

11:00 am   *Natural Products Drug Discovery at the National Cancer Institute*
Dr. Gordon Cragg (National Cancer Institute, NIH)

11:30 am   *The Camptothecin Experience: From Chinese Medicinal Plants to Potent Anti-cancer Drugs*
Dr. Stringner Sue Yang (National Cancer Institute, NIH)

12:00 Noon   Lunch

1:30 pm   *Collection of Medicinal Plants for New Drugs*
Dr. Jim Miller (Missouri Botanical Garden)

2:00 pm   *Mission and Projects of National Neutraceutical Center*
Dr. J. David Gangemi (Director, NNC)

2:30 pm   *Foods and Medicinal Plants as Sources of Health Enhancing Compounds*
Dr. Steven Bobzin (Monsanto Company)

3:00 pm   Coffee Break

3:30 pm   *Medicinal Plants in Jilin Province—Resources and Future Development*
Dr. Dalian Lu   (Deputy Secretary General, The People's Government of Jilin Province, China)

4:00 pm   *Development of Chinese Herbal Medicine in Hubei Province*
Mr. Lincheng Ma (Vice Mayor of Jinzhou City, Hubei)

4:30 pm   *Promotion of Herbal Medicine Industry with Scientific Technology in Native Place of Huatuo*
Mr. Zuojun Jiang (Vice Governor, Anhui Province, China)

THURSDAY, AUGUST 31, 2000

Scientific Session I: An Ancient Healing Art in a Modern World

Co-chairs: Dr. Ning-Sun Yang (Academia Sinica, Taipei)
Dr. Edward Croom (University of Mississippi)

9:00 am — *Botanicals: Searching for the Best of East and West*
Dr. Edward Croom (University of Mississippi)

9:30 am — *Evaluation Immune-modulation and Anti-tumor Activities in Specific Herbal Medicinal Plant Extracts*
Dr. Ning-Sun Yang (Academia Sinica, Taipei)

10:00 am — *Using Transcription Factor Based Assays to Study Herbal Products*
Dr. David Pasco (University of Mississippi)

10:30 am — Coffee Break

11:00 am — *On Quality Assessment of Chinese Patent Medicine*
Professor Peishan Xie (Guangzhou Institute for Drug Control)

11:30 am — *Modernization of Chinese Medicine Needs Five Finger Mountain and Golden Head Ring*
Dr. Paul But (Chinese University of Hong Kong)

Scientific Session II:
Chinese Herbal Medicine: A New Platform for Drug Discoveries

Co-chairs: Dr. Yuan Lin (Institute for Global Chinese Affairs)
Dr. Paul But (Chinese University Hong Kong)

1:30 pm — *Molecular Basis for Medicinal Actions of Androgens and Green Tea Epigallocatechin Gallate*
Dr. Shutsung Liao (University of Chicago)

2:00 pm — *Disease Control by the Compounds from Plants—Herbal Medicine in the 21st Century*
Dr. Ru-Chih Huang (The Johns Hopkins University)

2:30 pm	*Studies on Chemical Components and the Pharmacological Activity of Panax Ginseng Root* Professor Xianggao Li (Jilin Agricultural University)
3:00 pm	Coffee Break
3:30 pm	*Antimicrobial Activities in Rubricine: An Herbal Roots Extract of Arnebia Euchroma* Dr. Spencer Benson (University of Maryland)
4:00 pm	*A Practical Comprehensive Approach to Chinese Medicine Research* Dr. P.C. Leung (Chinese University of Hong Kong)

FRIDAY, SEPTEMBER 1, 2000

9:00 am   Special Presentation
"Back to Nature": The Alternative Paradigm for Drug Development
 Dr. Jasjit Bindra (Pfizer, Inc.)

9:30 am -12:00 Noon

FDA Forum: Regulation of Botanical Products

Presentation and Round Table Discussions
Moderator: Dr. Yuan Yuan Chiu (CDER, FDA)

Dr. Yuan Yuan Chiu, Director
(Office of New Drug Chemistry, CDER, FDA)
Dr. Chi-wan Chen, Director
(Division of New Drug Chemistry III, Office of New Drug Chemistry, CDER, FDA)

Dr. Ling Chin, Medical Officer
(Division of Over the Counter Drugs Evaluation, CDER, FDA)
Dr. Samuel W. Page, Director for Research
(Joint Institute for Food Safety and Applied Nutrition)
Dr. Lori Love, Director
(Clinical Research and Review Staff, CFSAN, FDA)
Dr. Joseph M. Betz, Vice President for Scientific and Technical Affairs
(American Herbal Products Association)
Dr. Freddie Ann Hoffman, Senior Director

(Medical and Clinical Affairs, Warner-Lambert Consumer
Health Care - R & D Pfizer, Inc.)

12:00 Noon   Lunch

Commercialization of TCM in the Global Market

Co-chairs: Dr. Robert Yuan (University of Maryland)
Ms.Peggy Brevoort (A. M. Todd Botanicals)

1:30 pm   *Global Markets for Botanical Products*
Ms. Peggy Brevoort (A. M. Todd Botanicals)

2:00 pm   *Opportunity and Challenges of Developing Medicinal Herbs*
Dr. Michael Chang (Pharmanex, Inc.)

2:30 pm   *TCM Botanical Drugs: Regulatory and Scientific Challenges*
Dr. Mark Edgar (Ancile Pharmaceuticals)

3:00 pm   Coffee Break

3:30 pm   *A New Approach to the Modernization of Chinese Herbal
Medicine*
Dr. Robert Yuan (Marco Polo Technologies, Inc.)

4:00 pm   *Commercialization of Chinese Herbal Medicine in a Global
Market*
Mr. Chung Guang Shen (Sun Ten Pharmaceuticals)

4:30 pm   *Development of Modern TCM in China Tianfa Group*
Jialong Gong (China Tianfa Group--Power 28 Co., Ltd.)

SATURDAY, SEPTEMBER 2, 2000

8:30 am   Exhibits and Poster Sessions (through 12:00 Noon)

INVITED SPEAKERS

BENSON, Spencer A.	Dept. of Cell Biology & Molecular Genetics, University of Maryland College Park, MA, 20742 Phone: 301-405-5478; Fax: 301-314-9489 Email: sb77@umail.umd.edu
BETZ, Joseph, M	Vice President of American Herbal Manufactures Association 8484 Georgia Ave., Suite 370 Silver Spring, MA 20910 Phone: 301-588-1171; Fax: 301-588-1174 Email: jbets@ahpa.org
BINDRA, Jasjit S.	Bldg 260, Room 2209, Pfizer Central Research, Eastern Point Road Groton, CT, 86340 Phone: 860-441-385; Fax: 860-441-5728 E-mail: jasjit_s_bindra@groton, pfizer.com
BOBZIN, Steven C.	Director, Natural Products and Analytical Chemistry, Galileo Laboratories 5301 Patrick Henry Drive, Santa Clara. CA 95054 Phone: 408-654-5830 xt.187 Fax: 408-369-5831 Email: sbobzin@GalileoLabs.com
BREVOORT, Peggy	2125 First Ave, #2503 Seattle, WA 98121 Phone: 206-239-0951; Fax: 541-271-2730 Email: pegbre@earthlink.net
BUT, Paul	Dept. of Biology, Chinese Univ. of Hong Kong, Rm E205, CMMRC, East Block Science Centre, Hong Kong Phone: 852-2609-6140 Fax: 852-2603-5248 Email: paulbut@cuhk.edu.hk

CHANG, Michael N.

Pharmanex Inc.
2000 Sierra Point Pkwy, Suite 701
Brisbane, CA 94005
Phone: 650-616-6300; Fax: 650-616-6399
Email: mchang@pharmanex.com

CHEN, Chi-Wan

Director, Division of New drug Chemistry
Center for Drug Evaluation and Research
U.S. Food and Drug Administration
5600 Fishers Lane
 Rockville, Maryland 20857
Phone: 301-827-2001; Fax: 301-827-2103
Email: chenc@cder.fda.gov

CHIN, Ling

Medical Officer, Division of Over-the –
Counter Drug Evaluation
Center for Drug Evaluation and Research
FDA, CDER/DOTCDP, HFD-560
5600 Fishers Lane
Rockville, Maryland 20857
Phone: 301-827-2265; Fax: 301-827-2315
Email: chinl@cder.fda.gov

CHIU, Yuan Yuan

Director, Office of New Drug Chemistry
Center for Drug Evaluation and Research
U.S. Food and Drug Administration
5600 Fishers Lane
 Rockville, Maryland 20857
Phone: 301-827-5918; Fax: 301-594-0746
Email: chiu@cder.fda.gov

CRAGG, Gordon M.

NCI-FCRDC Fairview Center
Suite 206, P.O.Box B
Frederick, Maryland, 21702-1201
Phone: 301-846-5387; Fax: 301-846-6178
Email:cragg@dtpax2.ncifcrf.gov

CROOM, Edward Jr.

Scientific and Regulatory Affairs Manager
Indena, USA East, Inc.
The Atrium, 80 East Route 4
Paramus, NJ 07405
Phone: 201-587-8883; Fax: 201-587-8048
Email: ed@indena.com

EDGAR, Mark T

Senior Vice President, Chemistry,
Manufacturing and control
Ancile Pharmaceuticals
10555 Science Center Drive, Suite B.
San Diego, CA 92121
Phone: 858-320-7899; Fax: 858-623-3396
Email: medgar@ancile.com

GAMGEMI, David J.

Executive Director
National Nutraceutical Center
5300 International Blvd
North Charleston, SC 29418
Phone: 843-760-3575; Fax: 843-760-4098
Email: gangemi@aticorp.org

GONG, Jialong

Chairman, Hubei Tianfa Group, Co., Ltd
11/F Tianfa Mansion
12 Jianghan N. RoadJingzhou
Hubei, China 434000
Phone:86-716-8560857
Fax: 86-716-8561690

HOFFMAN, Freddie Ann

Senior Director for Medical and Clinical
Affairs, Pfizer
201 Tabor Road, Morris Plains, NJ 07950
Phone: 973-385-2381; Fax: 973-631-7760
Email: freddie.hoffman@pfizer.com

HONG, Deyuan

Professor, Institute of Botany, Chinese
Academy of Sciences
20 Nanxincun, Xiangshan
Beijing 100093
People's Republic of China
Phone: 86-10- 62591431, 68342516
Fax: 86-10- 62590843
Email: hongdy@ns.ibcas.ac.cn

HUANG, Ru-Chih

Professor, Department of Biology
Johns Hopkins University, 144 Mudd Hall
3400 N. Charles Street
Baltimore, Maryland, 21218-2685
Phone: 410-516-7330; Fax: 410-516-5213

HUI, Yongzheng

Chairman and President
Shanghai Innovative Research Center of
Traditional Chinese Medicine
351 Guoshoujing Road
Zhangiang H-tech Park
Pudong New Area, Shanghai 201203,
People's Republic of China
Phone: 86-21-50801997
Fax: 86-21-3895 3247
Email: huiyongzheng@yahoo.com

JIANG, Zuojun

Vice Governor, Anhui Province
Wuhu People's Government
Wenhua Road, Wuhu, Anhui 241001
People's Republic of China
Phone: 86-553-3845764
Fax: 86-553-3852756
Email: whpcab@mail.ahwhptt.net.cn

LEUNG, P. C.

Professor, New Asia College
The Chinese University of Hong Kong
Shatin, New Territories, Hong Kong
Phone: (852) 2609-7609
Fax: (852) 2603-5418
Email: pingcleung@cuhk.edu.hk

LI, Xianggao

College of Chinese Medicine Material
Jilin Agricultural University
Changchun, Jilin, 130118
People's Republic of China
Phone: 86-431-4531589, 4531591
Fax: 86-431-4510971
Email: shihaiy@990.net

LIAO, Shutsung

Professor, Ben May Institute
University of Chicago
Box 424, 5841 S. Maryland
Chicago, IL, 60637
Phone: 773-702-6999; Fax: 773-834-1770
Email: sliao@huggins.bsd.uchicago.edu

LOVE, Lori A

Director, Clinical Research and Review
Office of Nutritional Products, Labeling and
Dietary Supplements, HFS 805
Center for Food Safety and Nutrition
Food and Drug Administration
200 C. Street, S. W. Washing D.C.
Phone: 202-205-4198; Fax: 202-205-3126
Email: lori.love@cfsan.fda.gov

LU, Lianda

Deputy Secretary General
The people's Government of Jilin Province
11 Xinfa Road, Changchun 120054
People's Republic of China
Phone: 86-431-8917091
Fax: 86-431-8917446

MA, Lincheng

Vice Mayor, City of Jingzhou
West Jianjing Road
Jingzhou, Hubei
People's Republic of China
Phone: 86-716-8252001

MILLER, James S.

Head, Applied Research Department
Missouri Botanical Garden
P.O.Box 299
St. Louis, Missouri, 63166
Phone: 314-577-5100; Fax: 314-577-9521
Email: james.miller@mobot.org

PAGE, Samuel W.

Director for Research
Joint Institute for Food Safety and Applied
Nutrition
200 C. Street, SW, Washington D. C.
Phoen: 202-205-4422; Fax: 202-205 4355
Email: swp@cfsan.fda.gov

PASCO, David S.

Professor, NCNPR, School of Pharmacy
University of Mississippi
University, MS 38677
Phone: 662-915-7130; Fax: 662-915-7062
Email:dpasco@sunset.backbone.
                olemiss.edu

RAVEN, Peter H.

Director, Missouri Botanical Garden
P.O.Box 299
St. Louis, MO, 63166-0299
Phone: 314-577-5111; Fax: 314-577-9595
Email: praven@nas.edu

SHEN, Chung Guan

Vice Chairman and Acting President
Sun Ten Pharmaceutical Co., Ltd.
7F, No. 200 Sec. 2, Chin-Shan S. Rd.
Taipei, Taiwan, Republic of China
Phone: 886-2-395-7070;
Fax: 886-2-321-3707

YANG, Ning-Sun

Director, Institute of Bioagricultural
Sciences Academia Sinica
Nankang, Taipei, Taiwan, 115
Republic of China
Phone: 886-2-26515910
Fax: 886-2-26515600
Email: nsyang@gate. sinica. edu. tw

YANG,, Stringner Sue

Chemist, Natural Products Branch
NCI-FCRDC, Fairview Center
Suite 206, P.O.Box B
Frederick, Maryland, 21702-1201
Phone: 301-846-5387; Fax: 301-846-6178
Email: stringners@dtpepn. nci. nih. gov

YUAN, Robert T.

Professor, Dept. of Cell Biology &
Molecular Genetics
University of Maryland
College Park, Maryland, 20742
Phone: 301-405-5436; Fax: 301-314-9491
Email: ry11@umail.umd.edu

XIE, Peishan

Guangzhou Institute for Drug Control
No. 23, Xizeng Street, Xicun
Guangzhou, Guangdong
People's Republic of China
Phone: (8620) 8652-9482
Fax: (8620) 8650-7466
Email: psxie@public.gz.cngb.com

SUBJECT INDEX

A

AIDS antiviral drug, 30
Aloe vera polysaccharide, 85
Androgen, 91
Apoptosis, 89
Antemisinin, 4
Anti-cancer, 86
Antigen-Presenting Cell (APC),
    76-77
Anti-inflammary, 84, 87
Authentication, 138

B

Bacteriocidal and bacteriostatic,
    115
Ben Cao Gang Mu, 59
Bioactive, 43, 44, 46, 47, 52, 53
Biological diversity, 50-52
Breast cancer, 92

C

Camptotheca, 147
Caompothecin analogs, 71
CD4+ T Cells, 80
Cell Cycle Arrest, 89
Chemical Fingerprinting, 183-184
Chemistry, Manufacturing &
    Control (CMC), 178, 181
Chinese Medicinal Material
    Research Center (CMMRC), 139
Chinese patent medicine (CPM),
    125
Chinese pharmacopoeia, 123

Clinical Trial Agreement (CTA),
    26
Clinical outcome studies, 194,
198,
    200
Combicine, 192-196
Computer informatics, 140
Cooperative Research and
    Development Agreement
    (CRADA), 26, 29
Cox-2, 86

D

Dendritic cell, 76-78, 82
Dietary Supplement, 168, 170
Dietary Supplement Health and
    Education Act (DESHA), 50,
    153, 155

E

Efficacy, 140
Epigallocatechin Gallate (EGCG),
    89, 91-95

F

Flora of China, 13, 15
Food & Drug Administration
    (FDA), 189
Food Library, 49-53
Food Library pyramid, 51
Functional food, 47

G

Ginseng, 97, 98, 100, 105, 109
Ginsenoside, 97, 98, 100-104
Global market, 145
Good agricultural practice (GAP),
 9, 54, 58, 135, 191
Good manufacture practice
(GMP),
 9, 51, 62, 135, 137
Global sales network (GSP), 1,
 9-10
Good Laboratory Practice (GLP),
 1, 10
GM-CSF, 76
Gram-positive bacteria, 114

H

Health care market, 44
High performance liquid
 chromatography (HPLC), 111,
 184, 123, 131
High throughout screening (HTS),
 53
Human genome project, 18

I

Immunomodulation, 74, 86
Index compounds, 81
Investigational New Drug
 Application (INDA), 154, 160

L

Lipopolysaccharide (LPS), 81

M

Material Transfer Agreement
 (MTA), 19, 24-25
Metasequoia, 14
Missouri Botanical Garden
 (MBG), 34-37

Minimum Inhibitory
 Concentration (MIC), 113
Micro algae polysaccharide, 86
Molecular nutrition, 52
Multiple drug resistance (MDR),
 109

N

National Cancer Institute (NCI),
 19, 20, 61
Naturacentical, 155
Natural Products Repository
 (RPD), 20, 24-26
Neuropeptides, 93
New Chemical Entity (NCE), 1, 7,
 177, 179
New Drug application (NDA), 161
NK, 80

O

Obesity, 93
Office of Dietary Supplements, 42
Over the counter drugs (OTC), 46-
 47, 158

P

Pedicularis, 15
Pharmecological activities, 97
Phytoestrogen, 147
Platelet agglutination, 104
Primulaceae, 15
Process validation, 184
Prostate tumor, 91

Q

Quality Control, 140

R

5 α-reductase, 90
Rheumatoid arthritis, 199
Rhododendron, 15

S

Safety, 140
Saponius, 5, 100
Serum cholesterol level, 46
Shikonin, 111
Silica gel thin layer
  chromatography (TLC), 127,
  129
Small Business Innovation
  Research (SBIR), 30
Stability, 185
Standardized TCM, 191-192
Synergy, 154

T

Topoisomerase, 67
Tumor necrosis factor
  (TNF-α), 79

U

US Department of Agriculture
  (USDA), 64

W

World Health Organization
  (WHO), 204